TRANSMISSION OF INFORMATION
IN THE OPTICAL WAVEBAND

TRANSMISSION OF INFORMATION IN THE OPTICAL WAVEBAND

L. G. KAZOVSKY

A HALSTED PRESS BOOK

JOHN WILEY & SONS, New York · Toronto
ISRAEL UNIVERSITIES PRESS, Jerusalem

Copyright © 1978 by Keter Publishing House Jerusalem Ltd.

All rights reserved

ISRAEL UNIVERSITIES PRESS
is a publishing division of
KETER PUBLISHING HOUSE JERUSALEM LTD.
P.O. Box 7145, Jerusalem, Israel

Published in the Western Hemisphere by
HALSTED PRESS, a division of
JOHN WILEY & SONS, INC., NEW YORK

Library of Congress Cataloging in Publication Data

Kazovsky, L.G.
   Transmission of information in the optical waveband.

   "A Halsted Press book."
   Includes bibliographical references and index.
     1. Laser communication systems. 2. Data transmission systems. I. Title.
TK5103.6.K37   621.38'0414   77-28102
ISBN 0-470-26294-X

Distributors for the U.K., Europe, Africa and the Middle East
JOHN WILEY & SONS, LTD., CHICHESTER

Distributors for Japan, Southeast Asia and India
TOPPAN COMPANY, LTD., TOKYO

Distributed in the rest of the world by
KETER PUBLISHING HOUSE JERUSALEM LTD.
IUP Cat. No. 22208–6
ISBN 0–7065–1582–X

Printed and bound by Keterpress Enterprises, Jerusalem
PRINTED IN ISRAEL

CONTENTS

Foreword   vii
Preface   ix
Introduction   xi

Chapter 1. EQUIPMENT FOR LASER COMMUNICATION SYSTEMS   1
  1.1. Lasers   1
  1.2. Optical Modulators   9
  1.3. Photodetectors   17
  1.4. Other Optical Instruments   33
    A. Filters   34
    B. Optical waveguides   36

Chapter 2. OPTICAL SIGNALS AND NOISE. OPTICAL COMMUNICATION CHANNEL   41
  2.1. Photon Statistics and Distribution of Detector Photocurrent   41
  2.2 Optical Communication Channels and Their Mathematical Description   46

Chapter 3. FEEDBACK IN DATA-TRANSMISSION SYSTEMS   55
  Introductory Remarks   55
  3.1. Introduction and Use of Feedback—Basic Techniques   56
  3.2. Mathematical Description of Feedback Systems   60

Chapter 4. NOISE IMMUNITY OF LASER COMMUNICATION SYSTEMS AND ITS IMPROVEMENT BY USE OF FEEDBACK — 65

Introductory Remarks — 65
4.1. Selection of Basic LCS. Noise Immunity of One-Way Systems — 66
4.2. Maximum Noise Immunity of Feedback Systems. Systems with Memory. Sequential Analysis — 75
   A. Passive spacing — 77
   B. System with active spacing — 89
4.3. Noise Immunity of Memoryless Systems with Feedback. Feedback Efficiency — 94
4.4. Optical Signal Fluctuations in a Turbulent Atmosphere. Feedback Efficiency — 101
4.5. Effect of Feedback Path and Limited Number of Repetitions on Feedback Efficiency — 106
Conclusions — 113

References — 115

Index — 119

FOREWORD

Although the principles of laser physics were first formulated in the early 1930s, it was not until 30 years later that the scientific community realized their implications and began to operate a laser system. Real technical applications evolved over the subsequent decade, and, since that time, lasers have been used for a variety of purposes: cutting, shaping, scientific "tools," medicine, military applications and, of course, communication.

Despite early recognition of the laser beam as a potential carrier of information, widely used systems were slow to develop. A major reason for this lag in practical application was that while communications was purely an engineering enterprise, laser physics remained purely a scientific one. Consequently, the problems of utilizing a laser as an information carrier were not fully understood by the physicists who undertook to solve them.

In the present monograph, laser communication is described and analyzed in a way that offers a base from which both engineers and physicists can build. In addition, this monograph can serve graduate students in the laser field by its delineation of the requirements of a laser system if it is to be employed in a practical communication system.

Even if the laser itself is operating ideally, there still exist basic problems in determining its most efficient use as a communications tool. The author presents the various alternatives and analyzes in depth the two-way feedback approach. Accuracy, noise immunity and energy requirements are considered.

Dr. N. Ben-Yosef
Jerusalem, 1977

PREFACE

The main aim of this book is to explore the possibilities offered by feedback to improve the noise immunity of laser data-transmission systems. The relevant material, concentrated in Chapter 4, is largely a systematic account of my own papers. To make the book accessible to a wider circle of readers, I have seen fit to begin with an exposition in Chapters 1 through 3 of additional material necessary for a full account of the subject. Those readers familiar with the material discussed may of course skip some of these chapters.

In order to analyze a problem, a scientist must first of all find a suitable model; he then simply analyzes this model. Sometimes we discover that the solution is also applicable to other problems, provided that the mathematical model is adequate for them. Such is the situation here. Although the terminology employed in this book relates mainly to laser communication systems, the formulas, conclusions and graphs can be readily applied to laser location, aerial reconnaissance and other data systems.

The book is intended for scientists, engineers and advanced university students specializing in optical data systems. An understanding of the material presupposes a knowledge of mathematical statistics and statistical decision theory at the university level.

INTRODUCTION

The first lasers were developed by Maiman [1] and the team of Javan, Bennett, and Herriot [2]. At that time — the early sixties — it was universally assumed that the laser "invasion" of technology would be rapid and effective. The tremendous amount of energy per unit frequency band in comparison with any other light source, the high spatial and temporal coherence of the radiation, the low divergence of the beam, small size — all these factors held promise of considerable advantages and wide applications, particularly in data systems. Applied laser research began in the fields of communications, radar, navigation, aerial reconnaissance, etc. In each of these fields the laser offered prospects of significant progress: in communication systems — sharp improvement in channel capacity and immunity to artificial noise, reduction of the probability of interception and interference to a negligible level; in radar and navigation systems — considerably higher precision and resolution, more complicated design of devices for effective generation of jamming; and so on. Moreover, any laser-based data system enjoys yet another important advantage over the analogous radio system: a large number of laser systems may operate at one wavelength in a limited range without appreciable cross-channel interference.

It is thus clear why research on the design of laser data systems began to expand in both scope and intensity. Within a short time, many technical obstacles were overcome; improved lasers [3], modulators [4, 5], photodetectors [6, 7] and other necessary optical and radio-engineering components [8, 9] were designed (a brief survey of laser technology will be given in Chapter 1). However, after considerable efforts had been expended, it became clear that laser systems possess a number of inherent shortcomings which hinder, restrict and sometimes even prevent their exploitation. These defects are mostly specific to laser systems operating in the open atmosphere, the very systems that had seemed most promising for wide application. The disadvantage, revealed by studies of laser propagation in the atmosphere, was the following. Turbulent inhomogeneities of the atmosphere result in marked fluctuations in the intensity of the received radiation and significantly modify the track of the beam; weather conditions (clouds, rain, snow) may weaken the intensity of the optical signal by several orders of magnitude and completely impair system performance [10, 11]. In addition, atmospheric turbulence, background noise level and weather conditions vary considerably from day to night, from season to season, and sometimes in the course of one operating session [12]. This necessitates the following design goals: 1) efficient and automatic systems for scanning, capture and tracking of the beam, and 2) reliable techniques for combatting such types of noise as energy fluctuations in the received beam and background noise of variable intensity. Without a solution to these two pivotal problems, progress in the design of laser data systems seems highly problematical.

The first problem seems to have found a satisfactory solution [13]. For example, in 1975 a communication system with clear-air beam propagation was produced: the Model 741 Laser Voice and Data Communicator [49]. The set-up time of the system is 15 minutes; it employs a gallium arsenide laser diode; the operating wavelength is 9000 Å (IR region), power 1.5 watt, reliability 99.99% at range 3 miles, pulse length 10 nsec. The system may be used without FCC licensing; the cost per unit terminal equipment is $2500.

The situation as regards the second problem is less favorable. The above example alone demonstrates that the reliability of the system is relatively low.

The search for a solution has led scientists and engineers to methods employed in traditional radio-frequency systems: adaptation (self-tuning) [11], coding, and recently also feedback [14, 15, 16].

This last approach, based on the principles of sequential analysis [17, 18], eliminates the need for complex coding systems and is essentially adaptive. It combines the advantages of the first two approaches and is sometimes free of their defects.

Let us consider in brief two examples of the use of sequential analysis in laser data systems.

1. *Communication system.* Consider a communication system designed for the transmission of two signals, 0 and 1. A conventional system might use a signal s_0 to transmit 0 and a signal s_1 for 1. A system with feedback will transmit the symbol i ($i = 0, 1$) with the help of k signals s_{ik}, as follows. To transmit i, the transmitter sends a signal s_{i1} to the receiver; if the receiver considers the received signal sufficiently reliable (according to some criterion), it makes a decision (0 or 1) and sends a signal along the feedback path, permitting transmission of the next symbol. But if the received signal is deemed unreliable, the feedback path is used to request additional data and the transmitter emits the signal s_{i2}. The process is then repeated.

2. *Aerial reconnaissance.* An aircraft flying over a region scans each element of the region using optical equipment, with the aim of detecting an object x differentiated from its surroundings by reflectivity or luminosity at a certain wavelength, or by some other optical parameter. In the traditional scheme, each element of the surface is scanned for the same length of time. But when sequential analysis is used the apparatus *itself* proceeds to the next element, once it is convinced that a decision as to the presence (or absence)

of x in the preceding element can be made with sufficient confidence.

As nonsequential algorithms are special cases of sequential analysis [17], an optimal sequential algorithm cannot possibly be inferior to a nonsequential one. As a rule, the use of a sequential procedure guarantees a considerable gain [16, 17], represented in the above examples by higher system reliability or by a reduction in the necessary transmitter power (see Chapter 4). In the second example, the gain may also be expressed in a faster decision rate or an increase in the altitude of the aircraft, an obvious advantage in military applications.

Of course, there is a price to be paid for these advantages. In the first example the cost is the need to introduce feedback. The pros and cons of this procedure and certain related topics will be discussed in Chapter 4.

It should be emphasized that the problems involved in the above two examples, though physically quite different, are almost the same from the mathematical point of view. Therefore, although the terminology employed in this book is mainly appropriate to the first example, the formulas, conclusions and graphs are readily carried over to the second, and also to other fields such as laser location and other data systems.

It should be clear from the above examples that the introduction of sequential-analysis principles into laser data systems is to a significant degree a universal method, successfully applicable in the solution of various specific problems.

At the same time, the sparse available data on the technique are spread over various journals, sometimes difficult of access. I have therefore undertaken to write this book, whose main topic may be formulated as follows: how can the noise immunity of a laser data-transmission system be improved by the use of feedback? As I have been working in this field for some time, my subjective views and inclinations have undoubtedly influenced the exposition. Thus, no mention is made of the problems of synchronization and the effect of feedback on the system data rate. On the other hand, I am sure that not all relevant material has been discussed or even mentioned. Nevertheless, I feel that "half a loaf is better than no bread." My main goal is merely to attract the attention of laser system specialists to sequential algorithms and feedback.

CHAPTER 1

EQUIPMENT FOR LASER COMMUNICATION SYSTEMS

Preview
Optical data systems make use of the following components: lasers, optical modulators, photodetectors, lenses, mirrors, deflectors (scanners), filters, optical quantum amplifiers, nonlinear crystals for frequency multiplication and transformation, axicons, optical waveguides and so on. This chapter will consider only lasers, modulators, photodetectors, filters and waveguides, which are the most essential for our purposes.

1.1. LASERS[1]

As witnessed by their name (laser = light amplification by stimulated emission of radiation), lasers are amplifiers of light, but they are used mainly as oscillators (the use of lasers as amplifiers is usually limited by internal noise).

We begin with a brief and simplified account of the operating principle of the laser. The basic component is an active medium whose particles (atoms or molecules) may be at one of three[2] energy levels (Figure 1.1). In the absence of external excitation, the atoms are at level 1, which has minimum energy. Upon application of excitation energy — pumping — the atoms jump to the broad level 3; to make the transition $1 \to 3$ possible, pumping must supply energy equal to the difference between levels 3 and 1.

---
[1] According to [3, 10, 19, 20, 46]; most of the figures and tables are taken from these sources.
[2] Four-level systems are also used.

2 Ch.1. EQUIPMENT FOR LASER COMMUNICATIONS

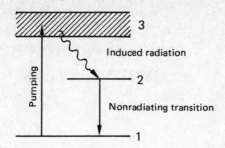

Fig. 1.1
Energy-level diagram of optical generator.

Thus, the population of level 3 (the number of atoms in stated 3) exceeds that of level 2. This process is known as population inversion.

In transition from the third level to the second, the atom radiates energy at frequency

$$f = \frac{E_3 - E_2}{h}, \qquad (1.1)$$

where E_3, E_2 are the energies of the respective levels and h is Planck's constant. If the transition is a result of the medium being irradiated at frequency f, it is said to be induced or stimulated; otherwise one speaks of a spontaneous transition. The probability of an induced transition is much higher than that of a spontaneous one.

Let us consider what happens when the active medium is placed between two mirrors forming a resonator (Figure 1.2). Radiation at frequency f, resulting from spontaneous

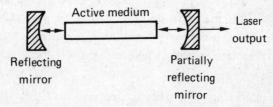

Fig. 1.2
Diagram of laser resonator.

transitions, causes induced transitions at the same frequency. The electromagnetic wave thus produced is amplified and travels along the resonator to the mirror. There it is reflected and sent back along the resonator, being amplified constantly. At the other end of the resonator the process is repeated.

We thus obtain a system with feedback which, as is well known, produces oscillation under suitable conditions. A steady state occurs when the gain achieved by the active medium is balanced by the losses. One reason for these losses is escape of energy through the partially transmitting mirror, which is of course a necessity to the user.

It follows from our model that the laser operates at a fixed frequency f, so that its oscillation spectrum is a δ-function. In practice, however, the oscillation spectrum of the laser medium is spread over about 1 Hz, owing to thermal agitation of atoms in the medium (Doppler broadening) and interatomic collisions. The precise oscillation frequency is determined by the resonator. If the oscillation spectrum of the medium includes the frequencies of several resonator modes and the laser gain in these modes exceeds the losses, the laser will oscillate in all these modes. This is precisely what happens in practice. Multimode oscillation of a laser affects the performance of the communication line in which it is incorporated [21, 22]. For this reason, special measures have been developed to suppress unwanted modes. To select transverse modes, one uses irises and other devices [23] to constrict the diameter of the resonator. For axial mode selection, one frequently inserts in the resonator a tilted Fabry–Perot etalon. Moreover, the oscillation frequency may be additionally stabilized [24].

Using all or some such measures in combination, one can achieve practically single-mode oscillation; this possibility will be utilized below (Chapter 2).

Lasers may be classified according to the type of medium: gas, semiconductor, solid-state (crystal), parametric, and dye lasers. The first three terms are self-explanatory; parametric lasers make use of nonlinear crystals, while dye lasers utilize liquids (dye solutions).

It seems that the most promising for high-capacity communication systems are gas and solid-state lasers [3], and we now discuss these two groups in greater detail.

The best known gas lasers are helium-neon (He-Ne), argon (Ar II) and carbon dioxide (CO_2) lasers. The basic characteristics of these lasers are listed in Table 1.1. (The letters E and H after RF (radio frequency) indicate the RF coupling methods used.) This table is reproduced from [3], as are the photographs in Figures 1.3, 1.4 and 1.5.

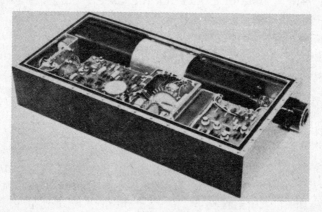

Fig. 1.3
3.5 mW He-Ne laser of integrated design. Power required 30 watts, dc voltage 24-32 volts (reproduced from [3] with kind permission of IEEE).

Fig. 1.4
Metal-ceramic capillary from BeO discharge tube of argon laser. Output of 5 watts from 25-cm active region (reproduced from [3] with kind permission of IEEE).

TABLE 1.1
Characteristics of Gas Lasers (reproduced from [3] with kind permission of IEEE)

	He–Ne	Ar II	CO_2
Wavelength (μ)	0.633	0.488	10.6
	1.15	0.515	9.6
	3.39	(0.45–0.53)	(9.6–10.8)
Medium	Glow discharge in He–Ne mix	Arc discharge in pure Ar	Glow discharge in CO_2–N_2–He
Current density (A/cm²)	0.05–0.5	100–2000	0.01–0.1
Excitation	dc, RF (E)	dc, RF (E, H)	dc, RF (E)
Power (at wavelength)	(0.633)	(multicolor)	(10.6)
Best laboratory (watts)	1	100	9000
Best commercial (mW)	100	20	1000
Lowest commercial (μW)	100	1	1
Efficiency (percent)	0.001–0.1	0.01–0.2	1–20

6 Ch.1. EQUIPMENT FOR LASER COMMUNICATIONS

Fig. 1.5
Early version of MOPA CO_2 laser system. Peak power 15 kW, average power 1.5 kW (reproduced from [3] with kind permission of IEEE).

He-Ne lasers are now highly developed. Specialists claim that their cost is approaching that of conventional (electronic) tubes. They are highly reliable (lifetime of the order of 10,000 hours); for this reason, many of the first operative communication channels used He-Ne lasers. However, the power of the commercial version is at most 100 mW, and that of the laboratory type almost 1 watt, so that designers have been stimulated to perfect the much more powerful Ar and CO_2 lasers. Particularly good results have been obtained with CO_2 lasers; according to recent reports, the power of both laboratory and commercial instruments may be raised more than one order of magnitude in comparison with the figures in Table 1.1. Progress in Ar-ion lasers is apparently being delayed by technical difficulties in handling the active medium. Design of CO_2 lasers has also encountered obstacles (for example, the difficulty of fabricating sufficiently efficient optics in the infrared), but their high efficiency and low quantum noise have stimulated development. Thus, there is a report [46] of a CO_2 laser with power 500 watt and efficiency more than 10%, designed for installation on aircraft.

Much attention has been devoted to optically pumped lasers utilizing yttrium aluminum garnet $Y_3Al_5O_{12}$ doped with Nd^{3+} ions. As a solid-state laser, this device is particularly attractive for military and space applications, where it is undesirable to have to deal with the flasks of gas and liquid lasers. Rods of Nd:YAlG 15 cm long 1 cm in diameter provide powers of up to 700 watts, and the few watts often required in laser communication systems are easily achieved (Figure 1.6). The laser illustrated in Figure 1.6 operates at several wavelengths in the infrared; the fundamental is 1.064 μ. The linewidth at room temperature is 180 GHz. If visible light is required, second-harmonic generation can be achieved by inserting a nonlinear crystal in the resonator.

More detailed characteristics of Nd:YAlG lasers are listed in Table 1.2.

The above examples suffice to demonstrate that modern optical communication systems have at their disposal lasers operating in both visible and infrared regions of the spectrum, at average powers ranging from 100 μW to 100 watts. These lasers can be operated in a single mode, at the cost of a loss of power by one or two orders of magnitude.

Fig. 1.6
Nd:YAlG laser with tungsten iodide pump; output of the order of 10 watts. On the right: power supply and closed-cycle water-cooling system (reproduced from [3] with kind permission of IEEE).

TABLE 1.2

Characteristics of Room Temperature Nd:YAIG Lasers (reproduced from [3] with kind permission of IEEE)

Mode of operation		$1.064\ \mu$	$0.532\ \mu$
High power CW multimode	(W-I)	150 watts, Efficiency 1.8 percent	
	(Kr)	100 watts, Efficiency 2.9 percent	
	(Kr)	250 watts, Efficiency 2.1 percent	
	(Kr)	750 watts, Efficiency 1.7 percent	
Low power multimode	(W-I)	10–15 watts, Efficiency 0.66–1.0 percent	4 watts
Single transverse mode		4–6 watts	
Single axial mode		2–3 watts	
Repetitively Q-switched			~ same as $1.064\ \mu$
Peak power		$10^3 P_{cw}$ $3 \times 10^2 P_{cw}$ $10^1 P_{cw}$	
Average power		$0.25 P_{cw}$ $0.73 P_{cw}$ P_{cw}	
Repetition rate (pps)		10^3 4×10^3 5×10^4	
Typical pulsewidth (μs)		0.25 0.61 2.0	
Repetitive cavity dumped			
Peak power		$1-2 \times 10^2 P_{cw}$ $1-2 \times 10^1 P_{cw}$ $1-2 P_{cw}$	
Average power		P_{cw} P_{cw} P_{cw}	
Repetition rate (pps)		10^5 10^6 10^7	
Typical pulsewidth (ns)		(50–100) (50–100) (50–100)	
Mode locked			~ same as $1.064\ \mu$
Average power		P_{cw}	
Peak power		~ $3 \times 10^2 P_{cw}$ ~ $3 \times 10^1 P_{cw}$	
Repetition rate (pps)		10^8 10^9	
Pulsewidth (ps)		30 30	

8 Ch.1. EQUIPMENT FOR LASER COMMUNICATIONS

We shall not discuss in any detail such topics as mode locking, Q-modulation and Q-switching; suffice it to say that all these methods of intra-resonator modulation yield pulsed lasers with peak power considerably exceeding the average. This explains their use in pulsed communication systems.

Injection lasers (which operate at room temperature) and chemical lasers are of recent appearance; therefore, although at present their characteristics do not seem very promising, it is too early to draw conclusions as to their application in communication systems. It may well be that the compactness of semiconductor lasers will in time force designers to ignore their shortcomings.

1.2. OPTICAL MODULATORS[3]

Methods for modulation of laser radiation may be divided into two groups: 1) direct (internal) modulation, 2) modulation outside the resonator.

In the first group, modulation is achieved by directly influencing the laser (for example, by varying the pumping power or inserting an optical modulator inside the resonator). In the second group the modulating effect is imposed on the radiation after it leaves the laser.

Direct modulation by change of pumping power is usually characterized by a significant time delay and high power consumption of the modulator. For this reason the method is not used in communication systems with high data rates (with the exception of semiconductor lasers). On the other hand, the use of a light modulator inside the resonator usually involves the same techniques as in modulation outside the laser.

Laser radiation may be modulated in regard to amplitude, frequency, phase and polarization. The modulators are

---

[3] Main sources of material, figures and tables: [4, 5, 10, 19, 25, 26].

based on the electrooptic, magnetooptic or acoustooptic effect; sometimes mechanical beam cut-off is also used.

Laser communication systems with high data rates require wideband modulators; the most commonly used techniques in the design of such modulators employ the electrooptic and/or acoustooptic effect. We shall examine the first of these effects in greater detail.

The electrooptic effect is observed in anisotropic dielectric media — crystals of various classes. The refractive index in such crystals depends on the propagation direction of the beam and on its polarization. In addition, it varies when an external electric field is impressed on the crystal; this phenomenon is known as the electrooptic effect.

The mechanism of the electrooptic effect is two-fold: 1) the electric field modifies electronic polarizability; 2) the field causes a lattice displacement, which in turn modifies electronic polarizability. For example, in the $LiNbO_3$ crystal the first component contributes no more than 10 percent of the overall effect.

In combination, these mechanisms may create a linear electrooptic effect (Pockels effect) and a quadratic effect (Kerr effect). In the first case the refractive index is a linear function of the applied field, in the second a quadratic function.

The optical properties of an anisotropic crystal are characterized by the optical indicatrix or index ellipsoid. The principal axes of the indicatrix usually lie along the crystal axes. The indicatrix is a surface whose properties are determined as follows.

Consider a plane electromagnetic wave propagating through the crystal. Imagine a plane through the center of the indicatrix, perpendicular to the phase-velocity vector of the wave. This plane intersects the indicatrix in an ellipse whose (directed) semiaxes may be treated as the polarization vectors of the normal modes. The lengths of the semiaxes determine the refractive indices for these modes.

The equation of the indicatrix is

$$\frac{x_1^2}{n_1^2}+\frac{x_2^2}{n_2^2}+\frac{x_3^2}{n_3^2}=1, \qquad (1.2)$$

where x_i are the coordinates and n_i the principal refractive indices ($i = 1, 2, 3$).

It can be shown that the Pockels effect represents the variation of n_i under an electric field:

$$\Delta n_i = \sum_k r_{ik} E_k, \qquad (1.3)$$

where E_k are the components of the applied field and r_{ik} the components of the linear electrooptic tensor ($i, k = 1, 2, 3$).

As an example, let us consider a crystal of KDP, once quite commonly used[4] (but now superseded by tantalite and lithium niobate crystals).

KDP is a crystal of the tetragonal system, with $\bar{4}2m$ symmetry at room temperature; the nonzero coefficients of the tensor r_{ik} are r_{41} and r_{63}.

If an electric field E_3 is applied to the crystal along the crystallographic c axis and the coordinate system is turned through 45° about the c axis relative to the crystallographic a and b axes, the equation of the indicatrix becomes:

$$\frac{x_1^2}{(n_o + \Delta n)^2} + \frac{x_2^2}{(n_o - \Delta n)^2} + \frac{x_3^2}{n_e^2} = 1, \qquad (1.4)$$

where

$$\Delta n = \frac{n_o^3}{2} r_{63} E_3. \qquad (1.5)$$

n_o is known as the refractive index of the ordinary wave, n_e that of the extraordinary wave. The form of the indicatrix is shown in Figure 1.7.

We now consider a wave propagating along the x_3 axis. If it is linearly polarized along the x_1 or x_2 axis, the only effect

---

[4] For detailed information on crystals, see [27].

12   Ch.1. EQUIPMENT FOR LASER COMMUNICATIONS

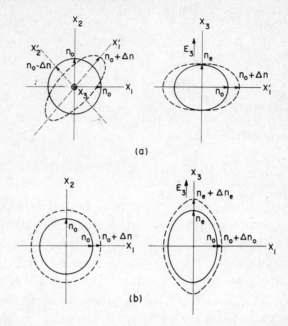

Fig. 1.7
Indicatrix before (solid curve) and after (dashed curve) application of field. (a) Indicatrix of KDP with field applied along x_3 axis; (b) indicatrix of LiTaO$_3$ with field applied along x_3 axis (reproduced from [4] with kind permission of IEEE).

of E_3 is to modify the refractive index for this wave, i.e., in the final analysis, the field causes phase modulation of the light. But if the wave is linearly polarized at an arbitrary nonzero angle to the x_1 and x_3 axes, it may be expressed as a superposition of normal modes polarized in the x_1 and x_3 directions. The effect of E_3 is then to make the refractive indices for the two normal modes differ from each other. Thus these modes are no longer in phase at the exit from the crystal, and their superposition produces an elliptically (not linearly) polarized wave: we have thus achieved polarization modulation. If the output wave is now passed through a polarizer (analyzer; see Figure 1.8), the result is amplitude modulation. The last type — frequency modulation — is

Sec. 1.2. OPTICAL MODULATORS   13

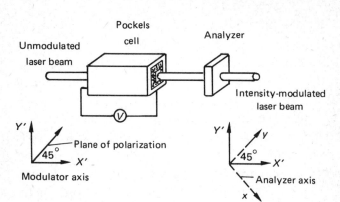

Fig. 1.8
Electrooptic amplitude modulator.

readily obtained by inserting a phase modulator in the laser optical resonator.

In other words, the electrooptic effect may be used to achieve any desired type of modulation. The modulators thus designed are convenient as to both size and power requirements. However, electrooptic crystals are highly sensitive to temperature and humidity, necessitating hermetic sealing, thermostats, partition of the modulator into two sections by a half-wave plate, and other special measures.

Modulators may be based either on the longitudinal electrooptic effect (electric field applied along the optical path) or on the transverse effect (applied field normal to the optical path). For a high index of modulation, the transverse electrooptic effect is preferable, as it yields a lower control voltage. The longitudinal electrooptic effect is used, for example, in eyepieces and objectives of controllable transparency: the thickness of a lens is much less than its diameter, and for the same voltage the longitudinal field is much stronger than the transverse field.

An essential factor, decisive for efficient performance of the modulator, is the mode of application of the modulating field. As long as the modulation frequency is relatively low, the modulating voltage cannot vary appreciably while the beam is traversing the crystal. In such cases one uses systems with lumped parameters, or (in the UHF range) resonant modulators. If the above condition is not fulfilled, modulation is a given frequency range implies use of the travelling-wave principle and the need to match the velocity of the modulating wave with the velocity of light in the crystal. This may be achieved by directly modifying the optical wave (zigzag modulator; the operating principle is clear from Figure 1.9). Frequently, however, the electromagnetic wave itself is modified, reducing its velocity by a special UHF structure (such as a strip waveguide).

When this is done, the modulation bandwidth may be limited 1) by velocity mismatch due to the design of the modulator itself, and 2) by the limited bandwidth of the waveguides, etc. that carry the modulating voltage.

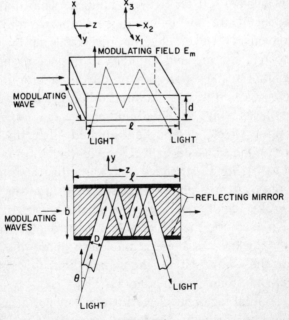

Fig. 1.9
Zigzag modulator (reproduced from [4] with kind permission of IEEE).

At the cost of a velocity mismatch at low frequencies (where it is not significant), the effect of the first factor may be reduced and the maximum bandwidth increased to 10 GHz [5]; the modulation bandwidth is then limited by the second factor alone.

When utilizing an electrooptic amplitude modulator, it must be remembered that the modulation characteristic (the output intensity of the light as a function of the applied voltage) is essentially nonlinear:

$$I_{out} = I_0 \sin^2(kv) \tag{1.6}$$

where I_{out} is the intensity of light at the modulator output, I_0 the intensity at the modulator input (it is assumed that the losses in the crystal are negligible), k is a factor depending on the design of the modulator and on the crystal used, v is the applied voltage.

Figure 1.10 illustrates the shape of the modulation characteristic and presents a diagram of I_{out} as a function of time for a sinusoidal modulating voltage. It is clear from the figure that without a dc bias the modulator nonlinearity is so significant that only the second harmonic of the modulating

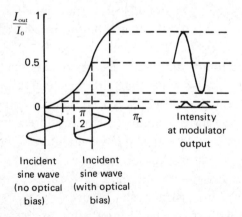

Fig. 1.10

Modulation characteristic of electrooptic modulator and output intensity I_{out} plotted against time (I_{out} plotted on vertical axis in relative units).

voltage appears at the output. In real modulators, therefore, one introduces a dc bias to reduce nonlinear distortions; the specific technique used may be either electrical or optical (by means of a quarter-wave plate). The second approach is preferable for practical purposes, as it does not require an additional dc source. It is clear, though, that the distortions are never eliminated, only reduced; the greater the desired depth of modulation, the larger the distortions. This correlation is highly undesirable in analog communication systems, where the goal is both large depth of modulation (to improve noise immunity) and small nonlinear distortions. This is one of the principal reasons for the superiority of discrete (digital) communication systems in the optical region. Such communication systems use the modulator as a shutter ("open" — "closed"); here, of course, there is no problem of nonlinear distortion. Instead, one has the problem of modulator contrast: even in the closed state the modulator passes a certain percentage of energy. To heighten contrast, two (or more) modulators may be connected in series.

The control voltages in modern modulators amount to a few volts, at most a few dozen; they may be lowered by using the modulator in a two-pass regime. The control powers are also not excessive. In modulator design, much attention is devoted to such side effects as nonuniform heating of the crystal by the optical beam and the control voltage. The latter is particularly important at high frequencies.

In conclusion, we note that although electrooptic modulators are now the most sophisticated and commonly used (and indeed we have examined them in more detail), there are situations in which other types of modulation are preferable. Acoustooptic modulators are now very popular; in the near infrared, magnetooptic modulation based on gallium-doped yttrium iron garnet may be used.

Materials with photoelastic interaction are also attracting attention nowadays, as their optical qualities are markedly superior. However, of all commercially accessible instruments, electrooptic and acoustooptic modulators seem the most suitable for external modulation.

When the communication system employs a semiconductor laser, direct modulation is used [28]; some relevant information on this topic may be found, e.g., in [29].

1.3. PHOTODETECTORS[5]

In this section we consider photodetectors in the narrow sense of the word, i.e., sensitive elements which convert an incoming optical signal into an electrical signal. At this point we shall not discuss the predetector elements (quantum amplifiers, frequency converters based on nonlinear crystals, filters, etc.) or the postdetector processing networks.

The requirements usually imposed on photodetectors for optical communication systems are the following:

1. High sensitivity on the frequency of the optical signal.
2. Sufficient bandwidth to accommodate the electrical signal.
3. Minimum internal noise.

In addition, such parameters as field of view, operating optical-frequency bandwidth, etc. are also important.

Photodetectors fall into two categories: thermal and quantum (photon). Thermal detectors (thermocouples, bolometers, etc.) are not used in optical data systems of the class we are considering, as they have a considerable rise time and therefore narrow bandwidth. We therefore turn immediately to quantum detectors. These in turn fall into four main groups:

1. Detectors utilizing the photoemission effect (also known as the external photoeffect), in which the incident light ejects electrons from the cathode of an electronic tube.

---

[5] Sources of material and most of the figures and tables: [6, 10, 19, 31].

The other three groups make use of semiconductors, in which the incident light results in changes in the concentration of charge carriers.

2. Photoresistors. These detectors utilize the photoconductivity effect, in which light incident on a semiconductor increases the concentration of charge carriers and thereby reduces the resistance of the detector, increasing the current in the circuit.

3. Valve detectors based on the photovoltaic effect. In these detectors light falls on the $p-n$ junction of a semiconductor (Figure 1.11). This generates electron-hole pairs, diffusion of which produces a potential difference at the junction. If an external bias is applied, the detector is called a photodiode. The design of a phototransistor, one of whose $p-n$ junctions is exposed to light, is similar.

4. Detectors utilizing the photoelectromagnetic effect. In these instruments, a magnetic field is applied to the semiconductor normal to the incoming beam. Then the electrons and holes generated by the light, diffusing into the semiconductor, are separated at the Hall angle by the magnetic field. The electrons are deflected to one surface of the semiconductor, the holes to the other. This produces a potential difference between the surfaces, proportional to the energy of the incident light.

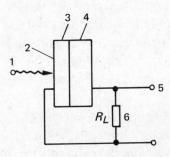

Fig. 1.11

Photovoltaic detector. 1) incident light, 2) optically transparent electrode, 3) p-field, 4) n-field, 5) output signal, 6) load resistor.

The first of these four groups, chronologically speaking, were photoemissive detectors.

The energy of an emitted electron in these detectors is

$$\mathscr{E}_e = h\nu - w, \qquad (1.7)$$

where ν is the frequency of the light, h Planck's constant and w the work function of the medium. It is evident from (1.7) that a necessary condition for emission of an electron (and so also for operation of the detector) is

$$\nu > \frac{w}{h}. \qquad (1.8)$$

However, even if this condition is fulfilled, some specific excited electron may fail to be emitted, owing to collisions with the atoms of the material. Thus the process as a whole is statistical, characterized by the quantum efficiency (sometimes also called quantum work function):

$$\eta = \frac{\bar{n}}{\bar{N}}, \qquad (1.9)$$

where $\bar{N}$ is the mean number of photons incident on the detector surface per unit time, $\bar{n}$ the mean number of emitted electrons (also per unit time).

The highest quantum efficiency is observed in semiconductor photodetectors; Figure 1.12 shows quantum efficiency plotted against wavelength.

Apart from quantum efficiency, other characteristics of the photodetector are: 1) dark current, 2) passband of the electrical signal, 3) internal noise, and 4) internal amplification.

Dark current is the current through the detector when there is no optical signal; it is caused by thermoemission, leakage current and extraneous fields.

The passband usually depends not on the photoelectric effect itself (which is practically instantaneous even in the UHF range), but on the subsequent mechanism of internal

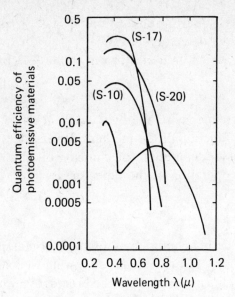

Fig. 1.12
Quantum efficiency of photoemissive materials.
S – 1: Ag–O–Cs
S – 10: Ag–Bi–O–Cs
S – 17: Cs–Sb
S – 20: Sb–K–Na–Cs

amplification (see below); the latter is also commonly the main source of internal noise.

A photodetector in which the electrons ejected from the cathode travel directly to the output is called a photoelement. In practice, however, the more common instrument is the photomultiplier, in which the electron beam is first focused through a series of special electrodes — dynodes (Figure 1.13). The potential of the first dynode is higher than that of the cathode; the potential of each successive dynode is higher than that of its predecessor; and finally the collector potential is higher than the potential of the last dynode.

Thus, for example, an electron ejected from the cathode is accelerated by the electric field and forces several (usually from three to five) electrons out of the dynode; this process is repeated at each dynode. As a result, the usual nine- or

Sec. 1.3. PHOTODETECTORS   21

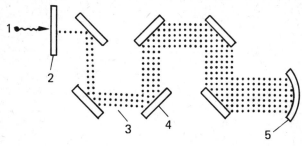

Fig. 1.13
Diagram of photomultiplier. 1) Incident light, 2) photoemissive surface (cathode), 3) secondary electrons, 4) dynodes, 5) collector.

ten-stage photomultiplier provides a gain of from 10^5 to 10^6. Since the process is statistical in nature, it generates the above-mentioned internal noise. In addition, electrons may have different transit times, and this, together with the intrinsic capacitance of the instrument, restricts the frequency characteristic.

To extend the photocathode, it may be combined with various UHF-structures. An example is the TW phototube of Figure 1.14. Here the photocathode is followed by a spiral traveling-wave tube, permitting operation at frequencies reaching from 10 to 20 GHz. There is a price to be paid, however: internal amplification is impossible.

Also used are static (Figure 1.15) and dynamic (Figure 1.16) crossed-field photomultipliers. Generally speaking, the two instruments rely on the same principle: the electric field accelerates the electrons, the magnetic field "twists" them

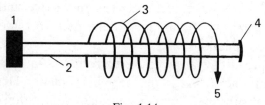

Fig. 1.14
Travelling-wave phototube. 1) Photocathode, 2) electron beam, 3) spiral delay structure, 4) electron collector, 5) UHF output.

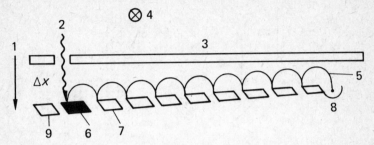

Fig. 1.15
Schematic of static crossed-field photomultiplier. 1) Direction of electric field E, 2) incident light, 3) passive electrode, $V = 0$, 4) magnetic field, 5) trajectory of photoelectron, 6) coaxial cable output, 7) first diode, $V = V_s + 2E\Delta x$, 8) photocathode, $V = -V_s + E\Delta x$, 9) cold electrode, $V = -V_s$.

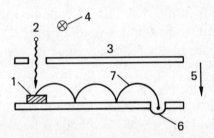

Fig. 1.16
Schematic of dynamic crossed-field photomultiplier. 1) Photocathode, 2) incident light, 3) passive electrode, 4) magnetic field, 5) RF electric field E, 6) coaxial cable output, 7) trajectory of electron.

and focuses them on the dynode. The difference is that the static instrument uses a discrete series of dynodes, while in the dynamic variant the dynodes are continuous, forming a compact active electrode. Between the active and passive electrodes an UHF pumping field performs the above-mentioned functions.

The spread of transit times is much less in crossed-field photomultipliers than in ordinary instruments. Hence they operate at substantially higher frequencies (in the gigahertz region), combining the high gain of the photomultiplier with the high frequency of the TW phototube.

Such sophisticated designs are highly efficient; however, according to [8], photoemissive detectors are serviceable only at sufficiently high optical frequencies. In the IR, therefore, one usually uses photoresistors and photodiodes (ordinary or avalanche). The avalanche diode differs from the ordinary solid-state diode in the same way as the photomultiplier differs from the photoelement. In other words, the avalanche photodiode possesses an internal multiplication mechanism analogous to that of the photomultiplier (though the physical principles on which it is based are different). The parameters of the above-mentioned three types of detector are listed in Tables 1.3, 1.4 and 1.5 (taken from [6]).

Selection of a suitable detector from this variety of instruments must be guided by specific calculations based on some optimality index. As an example, we consider the choice of a detector with maximum noise immunity for a laser communication system (LCS) with subcarriers.

Consider an optical binary-data transmission system with the following properties: 1) both transmitted message elements are equiprobable and equivalent; 2) the transmitter output is intensity-modulated by radio-frequency subcarriers; 3) in turn, each subcarrier is amplitude-modulated by its "own" information sequence of binary signals ("zero" and "one"); 4) the receiver is based on the block diagram of Figure 1.17.

The mean photocathode current is ([30]; see Figure 1.16)

$$\bar{I} = \frac{\eta e}{h\nu} \cdot P_s, \qquad (1.10)$$

where η is the quantum efficiency of the photodetector, $e = 1.6 \cdot 10^{-19}$ (coulombs) is the charge of the electron, $h = 6.62 \cdot 10^{-34}$ (joule · sec), ν is the optical carrier frequency, P_s is the optical signal power at the photosensitive element of the detector.

TABLE 1.3
Performance Characteristics of Photodiodes (reproduced from [6] with kind permission of IEEE)

Diode	Wavelength Range (μm)	Peak Efficiency (%) or Responsivity	Sensitive Area (cm²)	Capacitance (pF)	Series Resistance (Ω)	Response Time (seconds)	Dark Current	Operating Temperature (°K)	Comments
Silicon n⁺-p	0.4–1	40	2×10^{-5}	0.8 at −23 V	6	130 ps with 50-Ω load	50 pA at −10 V	300	avalanche photodiode
Silicon p-i-n	0.6328	>90	2×10^{-5}	<1	~1	100 ps with 50-Ω load	$<10^{-9}$ A at −40 V	300	optimized for 0.6328 Å
Silicon p-i-n	0.4–1.2	>90 at 0.9 μm >70 at 1.06 μm	5×10^{-2}	3 at −200 V 3 at −200 V	<1 <1	7 ns 7 ns	0.2 μA at −30 V	300	
Metal-i-nSi	0.38–0.8	>70	3×10^{-2}	15 at −100 V		10 ns with 50-Ω load	2×10^{-2} A at −6 V	300	
Au-nSi	0.6328	70	2			<500 ps		300	Schottky barrier, antireflection coating
PtSi-nSi	0.35–0.6	~40	2×10^{-5}	<1		120 ps		300	Schottky barrier, avalanche photodiode

Sec. 1.3. **PHOTODETECTORS** 25

Material	Range (μm)	η (%)						T (K)	Remarks
Ag-GaAs	<.36	50						300	
Ag-ZnS	<.35	70						300	
Au-ZnS	<.35	50						300	
Ge n^+-p	0.4–1.55	50	2×10^{-5}	0.8 at −16 V	<10	120 ps	$2 + 10^{-8}$	300	Germainum avalanche photodiode
Ge p-i-n	1–1.65	60 uncoated	2.5×10^{-5}	3		25 ns at 500 V		77	illumination entering from side
GaAs point contact	0.6328	40		0.027	30				
InAs p-n	0.5–3.5	>25	3.2×10^{-4}	3 at −5 V	12	$<10^{-6}$		77	
InSb p-n	0.4–5.5	>25	5×10^{-4}	7.1 at −0.2 V	18	5×10^{-6}		77	
InSb p-n	2–5.6		5×10^{-4}				1 MΩ shunt resistance	77	Reverse break down voltages 30 V
$Pb_{1-x}Sn_xTe$ 9.5 μm $x = 0.16$		45 V/W $\eta = 60$	4×10^{-3}			$\sim 10^{-9}$		77	shunt resistance $R_i = 10\,\Omega$
$Pb_{1-x}Sn_xSe$ 11.4 μm $x = 0.064$		3.5 V/W $\eta = 15$	7.8×10^{-3}			$\sim 10^{-9}$		77	shunt resistance $R_i = 2.5\,\Omega$
$Hg_{1-x}Cd_xTe$ 15 μm $x = 0.17$		$\eta \sim 10\text{–}30$	4×10^{-4}		8	$<3 \times 10^{-9}$		77	shunt resistance $R_i > 100\,\Omega$

TABLE 1.4
Characteristics of Avalanche Photodiodes (reproduced from [6] with kind permission of IEEE)

Diode	Wavelength Range (μm)	Sensitive Area (cm²)	Dark Current	Avalanche Breakdown Voltage (volts)	Maximum Gain	Multiplication Noise $i^2 \sim M^{2+x}$	Current Gain Bandwidth Product (GHz)	Capacitance (pF)	Series Resistance (Ω)
Silicon n^+–p	0.4–1	2×10^{-5}	50 pA at -10 V	23	10^4	$x \sim 0.5$	100	0.8 at -23 V	6
Silicon n^+–p π–p^+	0.5–1.1	2×10^{-3}		~88	200	$x \sim 0.4$	high		
Silicon n^+–i–p^+	0.5–1.1			200 to 2000		low	not very high		
PtSi–nSi	0.35–0.6	4×10^{-5}	~1 nA at -10 V	50	400	$x \sim 1$ for visible illumination	40 for UV excitation	<1	
Pt-GaAs	0.4–0.88			~60	>100	very low	>50		
Germanium n^+–p	0.4–1.55	2×10^{-5}	2×10^{-8} A -16 V and 300 K	16.8	250 at 300 K >10^4 at 80 K	$x \sim 1$	60	0.8 at -16 V	<10
Germanium n^+–p for 1.54				150					
InAs	0.5–3.5								
InSb at 77 K	0.5–5.5			a few	10	very low			

TABLE 1.5
Infrared Photoconductive Detectors (reproduced from [6] with kind permission of IEEE)

Material	Maximum Temperature for Background Limited Operation	Test Temperature (°K)	Long Wavelength Cutoff (50%) (μm)	Peak Wavelength λ_m (μm)	Absorption Coefficient (cm^{-1})	Quantum Efficiency η	Resistance (Ω)	$D^*_{\lambda_m}$ (cm·Hz$^{1/2}$/W)	Approximate Response Time (seconds)
InAs	110	195	3.6	3.3	$\sim 3 \times 10^3$			3×10^{11}	5×10^{-7}
InSb	60	77	5.6	5.3	$\sim 3 \times 10^3$	0.5–0.8	10^3–10^4	6×10^{10}	
Ge:Au	60	77	9	6	~ 2	0.2–0.3	4×10^5	3×10^9–10^{10}	3×10^{-8}
Ge:Au(Sb)		77	9	6			10^6	6×10^9 7×10^9 -4×10^{10}	1.6×10^{-9} 3×10^{-8}
Ge:Hg	35	4.2 27	14 14	11 10.5	~ 3 ~ 4	0.2–0.6 0.62	1–4 × 10^4 1.2×10^5	4×10^{10}	-10^{-9}
Ge:Hg(Sb)	35	4.2	14	11			5×10^5	1.8×10^{10}	3×10^{-10}–2×10^{-9} 3×10^{-10}–3×10^{-9}
Ge:Cu	17	4.2 20	27	23	~ 4	0.2–0.6	2×10^4	2–4×10^{10}	3×10^{-9}–10^{-8} 4×10^{-9}–1.3×10^{-7}
Ge:Cu(Sb)	17	4.2	27	23			2×10^5	2×10^{10}	$< 2.2 \times 10^{-9}$
Hg$_{1-x}$Cd$_x$Te $x = 0.2$		77	14	12	$\sim 10^3$	0.05–0.3	60–400	10^{10}	$< 10^{-6}$
Pb$_{1-x}$Sn$_x$Te $x = 0.17$–0.2		77 4.2	11 15	10 14	$\sim 10^4$		20–200 42 52	6×10^{10} 3×10^8 1.7×10^{10}	$< 4 \times 10^{-6}$ 1.5×10^{-8} 1.2×10^{-5}

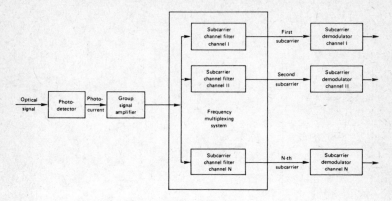

Fig. 1.17
Block diagram of optical communication line with frequency division multiplexing.

We assume the noise acting in the circuit to be concentrated at the detector output. To this end, we use the equivalent photodetector circuit of Figure 1.18, and characterize each type of noise by the variance of the appropriate stochastic process at the detector output (using the appropriate formulas from [30]):

1. Signal quantum noise: the variance is

$$\bar{i}_1^2 = 2eI_s \cdot \Delta f = \frac{2e^2\eta}{h\nu} \cdot P_s \cdot \Delta f, \qquad (1.11)$$

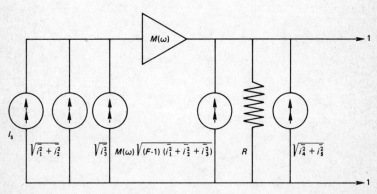

Fig. 1.18
Equivalent noise circuit of receiver.

where I_s is the signal photocurrent, P_s the signal power, Δf the frequency bandwidth.

2. Background noise:

$$\overline{i_2^2} = \frac{2e^2\eta}{h\nu} \cdot P_N \cdot \Delta f = \frac{2e^2\eta}{h\nu S} \cdot P_s \cdot \Delta f, \qquad (1.12)$$

where P_N is the power of the background noise, $S = P_s/P_N$ the signal/background noise ratio.

3. Noise due to dark current:

$$\overline{i_3^2} = 2eI_D \cdot \Delta f, \qquad (1.13)$$

where I_D is the dark current.

4. Photocurrent multiplication noise: In a photoreceiver with multiplication, the statistical nature of the process produces noise, which may be characterized by the multiplication-noise coefficient:

$$F = \frac{\text{signal/noise at input of multiplication section}}{\text{signal/noise at output of multiplication section}}. \qquad (1.14)$$

Incidentally, the multiplication process is frequency-dependent; this will be represented by the frequency characteristic $M(\omega)$ of the multiplication factor.

5. Thermal noise:

$$\overline{i_4^2} = \frac{4kT_F}{R} \cdot \Delta f, \qquad (1.15)$$

where k is the Boltzmann constant, T_F the equivalent noise temperature of the photodetector, R the equivalent noise resistance.

6. Thermal noise of the wideband group-signal preamplifier:

$$\overline{i_5^2} = \frac{4kT_A}{R} \cdot \Delta f, \qquad (1.16)$$

where T_A is the equivalent noise temperature of the preamplifier.

30 Ch.1. EQUIPMENT FOR LASER COMMUNICATIONS

Superposition of the noises (1) — (6) according to the scheme of Figure 1.17 yields the total variance of the noise:

$$\sigma^2 = \left\{ \left[\frac{2e^2\eta}{h\nu} \cdot P_s(1+S^{-1}) + 2eI_D \right] \cdot F \right.$$
$$\left. + \frac{4k(T_F+T_A)}{R \cdot |M(\omega)|^2} \right\} |M(\omega)|^2 \Delta f. \quad (1.17)$$

Moreover, it is readily seen from (1.10) that the amplitude of the signal photocurrent on the frequency subcarrier is

$$I_{sc} = I_s \cdot m \cdot |M(\omega)| = \frac{m\eta e}{h\nu} \cdot P_s \cdot |M(\omega)|, \quad (1.18)$$

where I_s is the signal photocurrent and P_s the signal power due to the dc component, and m is the depth of modulation of the light.

Now, knowing (1.17) and (1.18), we can find the signal/noise ratio at the subcarrier demodulator input (Figure 1.16):

$$h^2 = \frac{I_{sc}^2 \cdot \varphi_s^2(\tau)}{\sigma^2 \cdot \varphi_N}$$

$$= \frac{\left[\frac{m\eta e}{h\nu} \cdot P_s \cdot \varphi_s(\tau) \right]^2}{\left\{ \left[\frac{2e^2\eta}{h\nu} \cdot P_s \cdot (1+S^{-1}) + 2eI_D \right] \cdot F + \frac{4k(T_F+T_A)}{|M(\omega)|^2 \cdot R} \right\} \cdot \Delta f \cdot \varphi_N},$$

$$(1.19)$$

where $\varphi_s(\tau)$ represents the transmission of an elementary pulse of length τ through the channel filter, and φ_N the transmission of noise through the channel filter.

The functions $\varphi_s(\tau)$ and φ_N depend only on the type of frequency characteristic of the channel filter; they have been calculated once and for all for several basic types [45].

Thus, if the channel filter has an ideal U-shaped amplitude-frequency characteristic and a linear phase characteristic, then [45]

$$\varphi_s = \frac{2}{\pi} \cdot \text{Si}\left(\frac{\pi}{2} \cdot \Delta f \cdot \tau\right), \qquad \varphi_N = 1, \qquad (1.20)$$

where Si is the integral sine.

Another simple formula derived in [45] gives the error probability p ($p \ll 1$) as a function of h:

$$p = \exp\left(-\frac{h^2}{8}\right). \qquad (1.21)$$

Substituting (1.20) into (1.19) and (1.19) into (1.21) gives the error probability as a function of the optical signal energy. The function is shown in Figure 1.19 for four types of detector:
1) crossed-field photomultiplier,
2) photomultiplier-TW tube,
3) avalanche photodiode,
4) TW phototube.

Average characteristics of the above photodetectors are listed in Table 1.6. The last column of the table gives the optical signal power necessary to achieve $p = 10^{-6}$ with the detector in question. The calculation was done for a hypothetical LCS with the following parameters:

$\nu = 4.76 \cdot 10^{14}$ Hz (corresponding to the output frequency of an He-Ne laser);

$\tau = 8.75$ nsec (corresponding to a transmission rate of 114 Mbit/sec);

$\Delta f = 2/\tau \simeq 230$ MHz;
$S = 10$;
$m = 0.3$;
$T_A = 2{,}700°K$.

As may be seen from Figure 1.19, the noise immunity of an LCS is highest when a photomultiplier TW tube is used. Thus, to reach error probability 10^{-6} this detector requires a signal power of only 5.6 μW, whereas a crossed-field photomultiplier requires 11.0 μW, a TW phototube 11.7 μW, and an avalanche diode as much as 27 μW.

TABLE 1.6
Average Characteristics of Photodetectors

Type of detector	η	$|M(\omega)|^2 R$	F	$T_F(°K)$	I_D(amp)	P_s
Crossed-field photomultiplier	0.03	3.6×10^{10}	2	0	10^{-8}	11.0
Photomultiplier-TW-tube	0.05	5×10^7	1.7	0	~ 0	5.6
Avalanche photodiode	~ 1	200	4.8	300	0.22×10^{-6}	27
TW-phototube	0.05	5×10^5	1	0	0	11.7

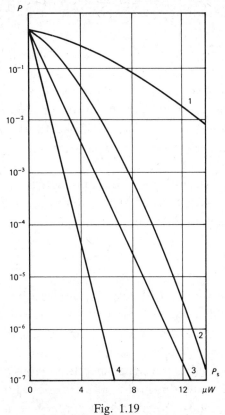

Fig. 1.19
Noise immunity of optical communication line with frequency division multiplexing, for different detectors: 1) avalanche photodiode, 2) TW phototube, 3) crossed-field photomultiplier, 4) photomultiplier-TW phototube.

In other words, the most efficient of the photodetectors we have examined is the photomultiplier-TW tube (using the least optical signal power to for the same error probability).

1.4. OTHER OPTICAL INSTRUMENTS

Besides the lasers, modulators and photodetectors already considered, laser communication systems also utilize following instruments:

1) Lenses and mirrors — in the receiving telescope and transmitter optics.

2) Deflecting elements (scanners) — in the systems for scanning, capture and tracking of the beam.

3) Filters — to suppress background noise in the receiver.

4) Special elements — quantum amplifiers, nonlinear crystals for optical-frequency multiplication and conversion, axicons, optical waveguides, and so on.

We shall discuss only filters and optical waveguides, as essential for our purposes; the reader interested in the other elements may consult, e.g., [8] and [10].

A. Filters

Light may be filtered with respect to optical frequency, space (stopped field of view), time (gating) and polarization.

Frequency filtration. Radio engineers know well that in order to achieve high noise immunity the frequency filter must be matched to the signal spectrum. However, it is often very difficult to do this in the optical region: the bandwidths of even the most sophisticated filters are usually much broader than the optical signal bandwidth. Therefore, the best optical filter from the standpoint of the LCS designer is frequently the narrowest of those available.

Available narrow-band filters include multilayer dielectric filters, electronically tunable acoustooptic filters, and Fabry–Perot interference filters. Recently developed Fabry–Perot filters provide optical bandwidths of 0.1 Å, and these are probably the most promising for optical communication purposes. A typical spectral characteristic is shown in Figure 1.20.

Polarization filtration. If the received optical signal is markedly linearly polarized or polarization-modulated, a polarization filter or polarizer should be used in the receiver. The quality of filtration is characterized by the ratio of the

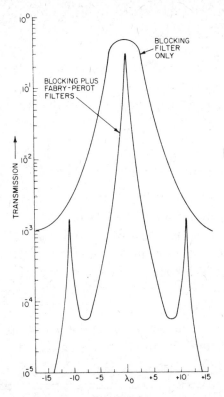

Fig. 1.20
Spectral characteristic of blocking filter combined with Fabry–Perot filter. Spectral characteristic of blocking filter shown for comparison ($\lambda_0 =$ wavelength at maximum transmission) (reproduced from [8] with kind permission of IEEE).

polarizer transmission factors for signals with mutually perpendicular directions of polarization. It may be improved by connecting two (or more) polarizers in series.

Temporal filtration (*gating*) may be introduced at the stage of postdetector processing; this type of filtration will therefore be covered by the discussion of optimal approaches to postdetector processing in the following chapters.

Spatial filtration. Theoretically speaking, an optimal optical filter can be synthesized on the basis of quantum-

36 Ch.1. EQUIPMENT FOR LASER COMMUNICATIONS

mechanical communication theory (a suitable approach was developed, e.g., in Helstrom [32]). If such a filter can be designed, it guarantees optimal filtration of all the above types. In practice, however, the signal is filtered with respect to each parameter separately; in particular, spatial filtering is achieved by simply using stops to limit the receiver's field of view.

B. OPTICAL WAVEGUIDES

Many different types of light-guide are available today [34, 46, 47]. They may be classified as follows:

1. Hollow waveguides, utilizing reflection of the optical beam from the walls of the waveguide (Figure 1.21a). The attenuation in such instruments is low, but the phase distortions are considerable and this limits their application to small-capacity systems.

2. Surface waveguides, utilizing the phenomenon of total internal reflection (Figure 1.21b). Single-mode waveguides

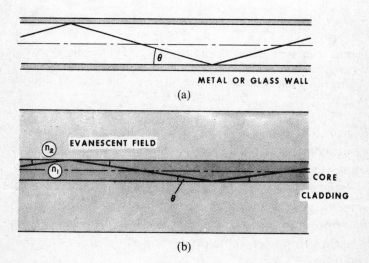

Fig. 1.21
Reflection waveguides using (a) reflecting walls, (b) total internal reflection (reproduced from [34] with knid permission of IEEE).

of this type hold promise of quite large bandwidths (a figure of 50 GHz is cited in [34]; even more optimistic estimates have been offered), but suffer from the defect of high attenuation. Nevertheless, the technology of these instruments is progressing rapidly, and reports have been published of specimens with attenuation of 10 db/km or even the possibility of approaching the theoretical limit of 2 db/km.

3. Graded index waveguides, in which the refractive index varies along the radius r; this class is a generalization of the surface waveguide, with more or less similar characteristics.

4. Film waveguides; the principle is illustrated in Figure 1.22.

5. Iris waveguides (Figure 1.23) ensure losses of less than 1 db/km, but require highly accurate alignment and permit practically no bends.

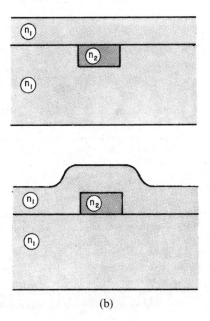

(b)

Fig. 1.22

Film waveguide with high-index channels (a) induced in, (b) deposited on lower-index substrate; both channels covered with a shielding layer (reproduced from [34] with kind permission of IEEE).

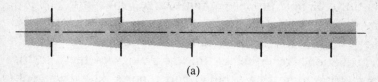

(a)

Fig. 1.23
Iris waveguide (reproduced from [34] with kind permission of IEEE).

6. Waveguides with solid lenses or mirrors (Figure 1.24); the attenuation is approximately 0.02 db per lens, yielding less than 0.5 db/km. Random fluctuations in these instruments may lead to large beam deflections and broadening, implying the need for automatic alignment schemes.

7. Gas-lens waveguides (Figure 1.25), as their name indicates, employ gas lenses instead of solid lenses. They guarantee exceptionally low attenuation and permit changes in the

Fig. 1.24
Lens waveguide (reproduced from [34] with kind permission of IEEE).

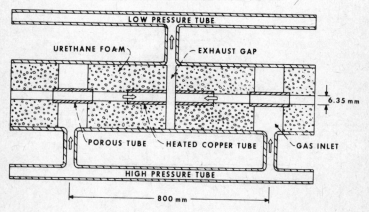

Fig. 1.25
Section of gas-lens waveguide, showing only one of the counterflow lenses (reproduced from [34] with kind permission of IEEE).

propagation direction of a beam, reaching bands with radius of curvature 500 to 600 meters.

A commercially available waveguide of class 2 was reported, e.g., in [47]. The diameter of the instrument is 0.5 cm, and its characteristics are: attenuation 20 db/km, weight 25 kg/km, minimum bend radius 2.5 cm, cost 70¢ per foot. It is claimed that the cost may be reduced to 10¢ per foot. The instrument transmits six independent channels with interchannel interference at most 60 db.

Waveguides of this type are already usable in working systems. Thus, in 1975 a waveguide communication system was produced with transmission rate 200 Mbit/sec at a distance of 100 m [47]. The system may also operate at a lower rate over larger distances — 200 m. It is intended for internal communication. The cost of the finished module is $400, that of the waveguide (including joints) $9 per meter.

There has recently been rapid progress in the development of fiber-optic links, which belong to classes 2 and 3 of those listed above [48, 49, 50, 51], and a reduction in the cost of the optical fiber should establish them firmly in the market.

CHAPTER 2

OPTICAL SIGNALS AND NOISE. OPTICAL COMMUNICATION CHANNEL

Preview
1. The photocurrent distribution in the detector depends on the structure of the incident optical flux. In some important cases, typical for optical data systems, the distribution is Poissonian. For example, the Poisson distribution is characteristic for the detection and identification of single-mode coherent signals on the background of multimode noise with small occupation number per mode.
2. After passing through a turbulent atmosphere, the laser beam shows fluctuations of intensity, phase, angle of arrival and beam dimensions. The intensity fluctuations are governed by a log-normal distribution, provided that the track is not too long and the aperture of the receiving antenna not too large.

2.1. PHOTON STATISTICS AND DISTRIBUTION OF DETECTOR PHOTOCURRENT

As in any problem of communication theory, analysis of an optical communication system requires, first and foremost, an adequate mathematical model of signal, noise and channel. We begin with the signal and the noise.

Present-day investigators employ two different approaches. The first specifies a direct description of the optical signal at the receiver input, while the second concentrates on describing the photocurrent in the detector(s).

The first approach is more general. Using quantum-mechanical methods, it enables one to achieve optimal synthesis of the circuits for both predetector and postdetector processing of the signal. However, it sometimes produces algorithms which are either unrealizable in principle or unrealizable at the present state-of-the-art. For this reason, the predetector processing circuit is frequently designed on the basis of "common sense" alone (e.g., one selects the optical filter with the narrowest bandwidth of those available). The second approach may then be applied.

As implied, the second approach presupposes that the predetector processing circuit is already available. The first problem to be solved is the following: knowing the properties of the optical signal at the detector input,[6] to devise a mathematical description of the photocurrent in the detector.

To this end, one can use Fermi's "golden rule" for the probability of a photon-electron transition in unit time [33]:

$$\frac{dP}{dt} \cong \frac{2\pi}{\hbar} |\langle f | H_I | i \rangle|^2 \rho(E_f), \tag{2.1}$$

where $\rho(E_f)$ is the energy density of the final states, and $\langle f | H_I | i \rangle$ the matrix element of the perturbation Hamiltonian between the initial and final states.

It follows from (2.1)[7] that the probability of an electron leaving the photodetector in a small time Δt is proportional to the energy $I(t)$ of the incident light:

$$P_1(t, t + \Delta t) = \alpha \cdot I(t) \cdot \Delta t, \tag{2.2}$$

where α is a proportionality factor. If Δt is sufficiently small, the probability of no electrons being ejected is

$$P_0(t, t + \Delta t) \cong 1 - \alpha \cdot I(t) \cdot \Delta t. \tag{2.3}$$

The probability that n electrons will be ejected in the time from 0 to $t + \Delta t$ is

$$P_n(0, t + \Delta t) = P_{n-1}(0, t) \cdot \alpha \cdot I(t) \cdot \Delta t + P_n(0, t) \\ \times [1 - \alpha \cdot I(t) \cdot \Delta t]. \tag{2.4}$$

---

[6] We shall assume that the detector is a quantum device; for the moment, we postulate that the internal noise is small in comparison with the quantum noise considered here.

[7] We are following [33].

Sec. 2.1. PHOTON STATISTICS AND DETECTOR PHOTOCURRENT

Subtracting $P_n(0,t)$ from both sides of (2.4), dividing by Δt and letting $\Delta t \to 0$, we obtain

$$\frac{dP_n}{dt} = \alpha \cdot I(t) \cdot P_{n-1}(t) - \alpha \cdot I(t) \cdot P_n(t). \tag{2.5}$$

The solution of this differential-difference equation is

$$P_n(t) = \frac{\left[\alpha \int_0^t I(t')dt'\right]^n \cdot \exp\left[-\alpha \int_0^t I(t')dt'\right]}{n!}. \tag{2.6}$$

Averaging (2.6) over the ensemble, we obtain

$$P_n(t_1 T) = \frac{\int_0^\infty (\alpha W)^n \cdot \exp(-\alpha W) \cdot P(W) \cdot dW}{n!}, \tag{2.7}$$

where $W = \int_t^{t+T} I(t')dt'$; $P(W)dW$ is the probability that W falls in the interval $(W, W + dW)$.

If the electromagnetic field varies significantly over the detector surface, we must also include the time-space envelope of the field; otherwise we may assume that the field is constant over the detector surface (point detector). In that case, Eq. (2.7) directly yields the photoelectron statistics.

Omitting the details, we proceed directly to the final formulas of photoelectron statistics in the cases of most interest for our purposes.

1. Known signal $S(t)$ incident on the photodetector, no noise wave. In this case the detector photocurrent is Poisson-distributed:

$$P(k) = \frac{(\alpha E_s)^k}{k!} \cdot \exp(-\alpha E_s), \tag{2.8}$$

where $P(k)$ is the probability that k photoelectrons will be emitted in time t, α is the quantum efficiency, and E_s the signal energy in time t.

2. Known signal incident on the photoreceiver, white noise of spectral density N_0. Before detection the light is

passed through a filter of bandwidth $2B$. The photocurrent is governed by the distribution

$$P(k) = \frac{(\alpha N_0)^k}{(1+\alpha N_0)^{k+2Bt+1}} \cdot \exp\left[\frac{-\alpha E_s}{1+\alpha N_0}\right] L_k^{2Bt}\left[\frac{-\alpha E_s}{\alpha N_0(1+\alpha N_0)}\right], \quad (2.9)$$

where $L_k^{2Bt}(\ )$ are the Laguerre polynomials.

3. Special case of 2 in which there is no signal. The photocurrent is governed by the negative binomial distribution:

$$P(k) = \binom{2Bt+k}{k}\left(\frac{1}{1+\alpha N_0}\right)^{2Bt+1}\left(\frac{\alpha N_0}{1+\alpha N_0}\right)^k. \quad (2.10)$$

4. Special case of 3, obtained by letting $2Bt \to 0$. The distribution of photoelectrons is

$$P(k) = \frac{(\alpha N_0)^k}{(1+\alpha N_0)^{k+1}} \quad (2.11)$$

5. Special case of 3 when $2Bt \gg 1$ and

$$\alpha N_0 \ll 1. \quad (2.12)$$

Here

$$P(k) = \frac{(\alpha 2BtN_0)^k}{k!} \cdot \exp(-\alpha 2BtN_0). \quad (2.13)$$

This distribution is formally the same as (2.8); in fact, this is the Poisson distribution with $\alpha \cdot 2Bt \cdot N_0$ the average noise photoelectron energy. Thus the Poisson distribution is typical both in case of a known signal and in the case of noise, provided condition (2.12) is fulfilled. The physical meaning of the condition is that for one mode there is much less than one noise photoelectron; this is usually valid in an optical channel.

5. Special case of 2 when $2Bt \to 0$. Then

$$P(k) = \frac{(\alpha N_0)^k}{(1+\alpha N_0)^{1+k}} \cdot \exp\left[-\frac{\alpha E_s}{1+\alpha N_0}\right] \cdot L_k\left(\frac{-\alpha E_s}{\alpha N_0(1+\alpha N_0)}\right). \quad (2.14)$$

Sec. 2.1. PHOTON STATISTICS AND DETECTOR PHOTOCURRENT 45

6. Special case of 2 when $2Bt \gg 1$ and $\alpha N_0 \ll 1$. In this case

$$P(k) = \frac{[\alpha(2BtN_0 + E_s)]^k}{k!} \cdot \exp[-\alpha(2BtN_0 + E_s)] \qquad (2.15)$$

is a Poisson distribution; a comparison of this formula with (2.8) and (2.13) readily shows that it represents the superposition of a Poisson-distributed signal and an independent Poisson-distributed noise.

It is evident from the above formulas that in many important problems the detector current is Poisson-distributed in all possible variants. We list a few examples:

1. Discrimination and detection of multimode noise-like signals with small mode populations, on the background of interference with analogous properties.

2. Discrimination and detection of single-mode coherent signals on the background of noise with small occupation number per mode.

3. Measurement of the level of a multimode signal (noise) with small occupation number per mode.

Bearing these and similar problems in mind, we shall generally assume in the sequel that the detector current in response to steady signals is Poisson-distributed.

To be alble to use this distribution in describing the performance of laser communication systems (LCS), we shall postulate that the laser output is single-mode and coherent. This is of course a simplification, disregarding a host of purely technical features of LCS and, in particular, ignoring the properties of the laser itself. For example, the operation of a laser involves fluctuation of gas discharge (in gas lasers), random mechanical vibrations of the resonator, beats between modes due to the nonlinearity of the lactive medium, and so on. As a result, the laser output radiation is of a highly complicated nature and any description of it as a single coherent mode is necessarily an approximation. Some questions relevant to the accuracy of this description have already been discussed elsewhere [38].

2.2. OPTICAL COMMUNICATION CHANNELS AND THEIR MATHEMATICAL DESCRIPTION

The beam produced by a laser transmitter may reach the receiver in either of two ways:

1. propagating along a special guide system (§1.4), or
2. propagating in free space between transmitter and receiver.

Transmission of optical energy through guide systems has several advantages: absence of background noise from external sources (the sun, the moon, etc.), independence of atmospheric condtions along the track, difficulty of interception, etc. However, it is by no means free of defects: high-quality optical waveguides are extremely costly, the technique cannot be used for communication with aircraft, fabrication of the track requires expensive construction work (sometimes involving excavation), organization of operative communication lines is slow; and so on.

Atmospheric channels also raise serious problems. Chief among these are the high background noise, scattering, cloudiness and atmospheric turbulence, all of which cause strong attenuation and fluctuations of intensity, propagation direction and dimensions of the beam. Moreover, all these factors vary in the course of a day, from day to night, from winter to summer, and sometimes during one brief communication session. The result is the need for considerable reserves of power, for effective optical systems to filter out the background, and for precision systems that guide, seek out, capture and track the beam. However, these shortcomings are often compensated by the more rapid organization of communications, the possibility of setting up a link through space inaccessible to waveguides (enemy territory, water barriers, etc.), and the elimination of costly and often unreliable optical waveguides.

Sec. 2.2. OPTICAL COMMUNICATION CHANNELS

In the sequel we shall be concerned with open-air propagation systems.

As stated previously, compensation for fluctuations in the angle of arrival is the task of special guide systems. There remain problems arising from fluctuations in the intensity and phase of the optical wave. These fluctuations are usually characterized by the following parameters:

a) the variance of the logarithm of the intensity (log-intensity),

b) the covariance of the logarithm of the amplitude (log-amplitude), and

c) the structure function of the phase.

Formulas for these parameters appear in Tables 2.1, 2.2 and 2.3 (from [12]).

It should be emphasized that the logarithm of the intensity of the received radiation is usually normally distributed — this has been shown in many theoretical and experimental works (e.g., [25, 45]), on the assumption that the track is not too long and the apertures of the receiving antennas not too large. Hence the distribution of the intensity fluctuations is uniquely determined by the expectation $\bar{E}_s$ of the intensity and the variance σ^2 of log-intensity (which may be determined from the formulas in Table 2.1):

$$w(E_s) = \frac{1}{\sigma\sqrt{2\pi} \cdot E_s} \cdot \exp\left[-\frac{1}{2\sigma^2} \cdot \ln^2\left(\frac{E_s}{\bar{E}_s} \cdot e^{\sigma^2/2}\right)\right], \quad (2.16)$$

where $w(E_s)$ is the probability density of the signal intensity E_s.

Equation (2.16) defines the log-normal distribution. We shall use it in calculating the noise immunity of LCS operating in a turbulent atmosphere.

TABLE 2.1
Variance of Log-Intensity (reproduced from [12] with kind permission of IEEE)

Case	General Random Medium	Locally Isotropic Random Medium*
Plane wave, homogeneous medium	$\sigma_{\ln I}^2 = 8\pi^2 k^2 L \int_0^\infty \left(1 - \frac{k}{\kappa^2 L}\sin\frac{\kappa^2 L}{k}\right)\Phi_n(\kappa)\kappa\,d\kappa$	$\sigma_{\ln I}^2 = 1.23 C_n^2 k^{7/6} L^{11/6},\quad l_0^2/\lambda \ll L$
	$\sigma_{\ln I}^2 = \frac{4}{3}\pi^2 L^3 \int_0^\infty \Phi_n(\kappa)\kappa^5\,d\kappa,\quad L \ll l_0^2/\lambda$	$\sigma_{\ln I}^2 = 12.8 C_n^2 L^3 l_0^{-7/3}$
Plane wave, smoothly varying medium	$\sigma_{\ln I}^2 = 16\pi^2 k^2 \int_0^\infty d\eta\, C_n^2(\eta) \int_0^\infty d\kappa\,\kappa\,\Phi_n^{(0)}(\kappa)$ $\times \sin^2\!\left[\frac{\kappa^2(L-\eta)}{2k}\right]$	$\sigma_{\ln I}^2 = 2.24 k^{7/6} \int_0^L C_n^2(\eta)(L-\eta)^{5/6}\,d\eta,\quad l_0^2/\lambda \ll L$ $\sigma_{\ln I}^2 = 38.4 l_0^{-7/3} \int_0^L C_n^2(\eta)(L-\eta)^2\,d\eta,\quad L \ll l_0^2/\lambda$
Spherical wave, homogeneous medium	$\sigma_{\ln I}^2 = 8\pi^2 k^2 L \int_0^\infty \left\{1 - \left(\frac{2\pi k}{\kappa^2 L}\right)^{1/2}\left[\cos\!\left(\frac{\kappa^2 L}{4k}\right) C\!\left(\left(\frac{\kappa^2 L}{2\pi k}\right)^{1/2}\right)\right.\right.$ $\left.\left. + \sin\!\left(\frac{\kappa^2 L}{4k}\right) S\!\left(\left(\frac{\kappa^2 L}{2\pi k}\right)^{1/2}\right)\right]\right\} \Phi_n(\kappa)\kappa\,d\kappa^{\dagger}$	$\sigma_{\ln I}^2 = 0.50 C_n^2 k^{7/6} L^{11/6},\quad l_0^2/\lambda \ll L$
	$\sigma_{\ln I}^2 = \frac{2}{15}\pi^2 L^3 \int_0^\infty \Phi_n(\kappa)\kappa^5\,d\kappa,\quad L \ll l_0^2/\lambda$	$\sigma_{\ln I}^2 = 1.28 C_n^2 L^3 l_0^{-7/3},\quad L \ll l_0^2/\lambda$
Spherical wave, smoothly varying medium	$\sigma_{\ln I}^2 = 16\pi^2 k^2 \int_0^L d\eta\, C_n^2(\eta) \int_0^\infty d\kappa\,\kappa\,\Phi_n^{(0)}(\kappa)$ $\times \sin^2\!\left[\frac{\kappa^2 \eta(L-\eta)}{2kL}\right]$	$\sigma_{\ln I}^2 = 0.56 k^{7/6} \int_0^L C_n^2(\eta)\left(\frac{\eta}{L}\right)^{5/6}(L-\eta)^{5/6}\,d\eta,$ $l_0^2/\lambda \ll L$ $\sigma_{\ln I}^2 = 46.7 l_0^{-7/3} \int_0^L C_n^2(\eta)\left(\frac{\eta}{L}\right)^2 (L-\eta)^2\,d\eta,$

medium

$$-\cos\left(\frac{\gamma_1(L-\eta)}{k}\kappa^2\right)\Bigg]\Bigg||H|^2\Phi_n(\kappa)$$

$$\cdot\left\{\frac{3}{8}{}_1F_1\left(-\frac{5}{6},1;\frac{2\rho^2}{W_0^2[1+(\alpha_1 L)^2]}\right)-g(\alpha_1 L)\right\}^{\ddagger}$$

where

$$\gamma_1=\frac{1-\alpha_2 L+\{(\alpha_1^2+\alpha_2^2)L-\alpha_2\}\eta}{(1-\alpha_2 L)^2+(\alpha_1 L)^2}\quad\gamma_2=\frac{\alpha_1(L-\eta)}{(1-\alpha_2 L)^2+(\alpha_1 L)^2}$$

$$\alpha_1=\lambda/(\pi W_0^2)\qquad \alpha_2=1/R_0$$

$$\alpha=\alpha_1+i\alpha_2\qquad \rho=(y^2+z^2)^{1/2}=\text{distance from beam center}$$

$$|H|^2=\kappa^2\exp\left\{-\frac{\gamma_2(L-\eta)}{k}\kappa^2\right\}$$

where

$${}_1F_1(a,c;z)=\Phi(a,c;z)\text{ is the Kummer function}$$

$$g(\alpha_1 L)=\int_0^1\text{Re}\left[(1-x)^2-i\frac{[1+(\alpha_1 L)^2]x[1-x]}{\alpha_1 L}\right]^{5/6}dx$$

$$=\text{Re}\left\{\frac{6}{11}\left(\frac{1+i\alpha_1 L}{i\alpha_1 L}\right)^{5/6}{}_2F_1\left(-\frac{5}{6},1;\frac{17}{6};i\alpha_1 L\right)\right\}$$

$${}_2F_1(a,b;c;z)\text{ is a hypergeometric function}$$

$$\sigma_{\ln I}^2=4\pi^2(0.033)[-\Gamma(-5/6)]k^{7/6}L^{11/6}\int_0^1 d\eta C_n^2(\eta)G_A(\eta)$$

Beam wave, smoothly $\sigma_{\ln I}^2=8\pi^2\int_0^L d\eta C_n^2(\eta)\int_0^\infty d\kappa\kappa\left[J_0(i2\gamma_2\kappa\rho)\right.$
varying medium

$$\left.-\cos\left(\frac{\gamma_1(L-\eta)}{k}\kappa^2\right)\right]\Bigg||H|^2\Phi_n^{(0)}(\kappa)\quad\text{where}$$

$$G_A(x)=\text{Re}[(\gamma_2+i\gamma_1)(1-x)+S_m^2]^{5/6}-[\gamma_2(1-x)+S_m^2]^{5/6}$$

$$S_m=(k/L)^{1/2}l_0/5.92\|$$

---

* For the locally isotropic random medium $\Phi_n(\kappa)=0.033\,C_n^2(x)\kappa^{-11/3}\exp(-\kappa^2/\kappa_m^2)$. For the homogeneous medium $C_n^2(x)$ is a constant independent of x.

‡ Receiver a distance ρ distance ρ from beam center, beam collimated ($R_0=\infty$).

‖ Receiver at beam center, any value of R_0.

TABLE 2.2
Covariance of Log-Amplitude (reproduced from [12] with kind permission of IEEE)

Case	General Random Medium	Locally Isotropic Random Medium
Plane wave, homogeneous medium	$B_x(\rho) = 2\pi^2 k^2 L \int_0^\infty \left(1 - \frac{k}{\kappa^2 L}\sin\frac{\kappa^2 L}{k}\right)\Phi_n(\kappa) J_0(\kappa\rho)\kappa\, d\kappa$, $\quad L \ll l_0^2/\lambda$ $B_x(\rho) = \frac{1}{3}\pi^2 L^3 \int_0^\infty \kappa^4 \Phi_n(\kappa) J_0(\kappa\rho)\kappa\, d\kappa$, $\quad L \ll l_0^2/\lambda$	For $l_0^2/\lambda \ll L$, $b_x(\rho) = \begin{cases} 1 - 12.3 \dfrac{\rho^2}{(\lambda L)^{5/6} l_0^{1/3}}, & \rho \ll l_0 \\[4pt] 1 - 2.36\left(\dfrac{k\rho^2}{L}\right)^{5/6} + 1.71\dfrac{k\rho^2}{L} - 0.024\left(\dfrac{k\rho^2}{L}\right)^2 \\[4pt] +\cdots, & l_0 \ll \rho \ll (\lambda L)^{1/2} \\[4pt] -0.0242\left(\dfrac{k\rho^2}{4L}\right)^{-7/6}, & (\lambda L)^{1/2} \ll \rho * \end{cases}$ $b_x(\rho) = {}_1F_1(7/6, 1, -\kappa_m^2 \rho^2/4),\ L \ll l_0^2/\lambda *$
Plane wave, smoothly varying medium	$B_x(\rho) = 4\pi^2 k^2 \int_0^L d\eta\, C_n^2(\eta)\int_0^\infty d\kappa\,\kappa J_0(\kappa\rho)\Phi_n^{(0)}(\kappa)$ $\times \sin^2\left[\dfrac{\kappa^2(L-\eta)}{2k}\right]$	$B_x(\rho) = 4\pi^2 \cdot 0.033\, k^2 \int_0^L d\eta\, C_n^2(\eta) \int_0^\infty d\kappa\,\kappa J_0(\kappa\rho)$ $\times \kappa^{-11/3}\exp(-\kappa^2/\kappa_m^2)\sin^2\left[\dfrac{\kappa^2(L-\eta)}{2k}\right]*$ $B_x(\rho) = 9.6 l_0^{-7/3}\,{}_1F_1(7/6, 1, -\kappa_m^2\rho^2/4)\int_0^L C_n^2(\eta)(L-\eta)^2 d\eta,$ $L \ll l_0^2/\lambda *$
Spherical wave,	$B_x(\rho) = 4\pi^2 k^2 \int_0^\infty d\kappa\,\kappa\, \Phi_n(\kappa) \int_0^L d\eta\, J_0\!\left(\dfrac{\rho\kappa\eta}{L}\right)$	$b_x(\rho) = 1 - 2.2\left(\dfrac{k\rho^2}{L}\right)^{5/6} + 1.71\dfrac{k\rho^2}{L} + 0.050\left(\dfrac{k\rho^2}{L}\right)^{17/6}$

Spherical wave, smoothly varying medium

$$B_x(\rho) = 4\pi^2 k^2 \int_0^\infty d\kappa\kappa \Phi_n^{(0)}(\kappa) \int_0^L d\eta C_n^2(\eta) J_0\left(\frac{\rho\kappa\eta}{L}\right)$$

$$\times \sin^2\left[\frac{\kappa^2\eta(L-\eta)}{2kL}\right]$$

$$B_x(\rho) = 4\pi^2 0.033 k^2 \int_0^L d\eta C_n^2(\eta) \int_0^\infty d\kappa\kappa J_0\left(\frac{\rho\kappa\eta}{L}\right)$$

$$\times \kappa^{-11/3} \exp(-\kappa^2/\kappa_m^2) \sin^2\left[\frac{\kappa^2\eta(L-\eta)}{2kL}\right]$$

$l_0 \ll \lambda \ll L, \quad l_0 \ll \rho \ll (\lambda L)^{1/2}$

Beam wave, homogeneous medium

$$B_x(\rho) = \pi^2 \int_0^L d\eta \int_0^\infty d\kappa\kappa \{[J_0(\kappa P) + J_0(\kappa P^*)]|H|^2$$
$$+ J_0(\kappa Q) H^2 + J_0(\kappa Q^*) H^{*2}\} \Phi_n(\kappa)^{\ddagger}$$

No cases worked out.

where

$P^2 = [(\gamma_1 y_d - i2\gamma_2 y_c)^2 + (\gamma_1 z_d - i2\gamma_2 z_c)^2]$

$Q^2 = (\gamma y_d)^2 + (\gamma z_d)^2$

$y_d = y_1 - y_2, \quad z_d = z_1 - z_2$

$y_c = \tfrac{1}{2}(y_1 + y_2), \quad z_c = \tfrac{1}{2}(z_1 + z_2)$

$H^2 = -k^2 \exp\left\{-i\frac{\gamma(L-\eta)}{k}\kappa^2\right\}$

$\gamma = \gamma_1 - i\gamma_2$

Beam wave, smoothly varying medium

Let $\Phi_n(\kappa) = C_n^2(\eta) \Phi_n^{(0)}(\kappa)$ in previous formula.

No cases worked out.

---

* $\Phi_n(\kappa) = 0.033 \, C_n^2(x) \kappa^{-11/3} \exp(-\kappa^2/\kappa_m^2)$. For homogeneous medium, $C_n^2(x)$ = constant.

† $\Phi_n(\kappa) = 0.033 \, C_n^2(x) \kappa^{-11/3}$.

‡ $\rho_1 = (y_1, z_1), \rho_2 = (y_2, z_2)$ are the observation points referred to the center of the beam. Other symbols defined in Table 2.1.

TABLE 2.3
Structure Function of Phase (reproduced from [12] with kind permission of IEEE)

Case	General Random Medium	Locally Isotropic Random Medium*
Plane wave, homogeneous medium	$D_s(\rho) = 4\pi^2 k^2 L \int_0^\infty [1 - J_0(\kappa\rho)]$ $\times \left[1 + \frac{k}{\kappa^2 L}\sin\frac{\kappa^2 L}{k}\right]\Phi_n(\kappa)\kappa\, d\kappa$ $D_s(\rho) = 8\pi^2 k^2 L \int_0^\infty [1 - J_0(\kappa\rho)]\Phi_n(\kappa)\kappa\, d\kappa,$ $L \ll l_0^2/\lambda$	For $l_0^2/\lambda \ll L$, $D_s(\rho) = \begin{cases} 1.64 C_n^2 k^2 L l_0^{-1/3}\left[1 + 0.87\left(\dfrac{l_0^2}{\lambda L}\right)^{1/6}\right]\rho^{2\dagger} \\ +\cdots,\ \rho \ll l_0 \\ 1.46 C_n^2 k^2 L\rho^{5/3},\ l_0 \ll \rho \ll (\lambda L)^{1/2} \\ 2.92 C_n^2 k^2 L\rho^{5/3},\ (\lambda L)^{1/2} \ll \rho \end{cases}$ For $L \ll l_0^2/\lambda$, $D_s(\rho) = 0.132\pi^2\dfrac{6}{5}\Gamma\left(\dfrac{1}{6}\right) k^2 LC_n^2 \kappa_m^{-5/3}$ $\times\left[{}_1F_1\left(-\dfrac{5}{6}, 1, -\dfrac{\kappa_m^2\rho^2}{4}\right) - 1\right]$
Plane wave, smoothly varying medium	$D_s(\rho) = 8\pi^2 k^2 \int_0^L d\eta\, C_n^2(\eta)\int_0^\infty d\kappa\kappa[1 - J_0(\kappa\rho)]$ $\times \Phi_n^{(0)}(\kappa)\cos^2\left[\dfrac{\kappa^2(L-\eta)}{2k}\right]$	$D_s(\rho) = 8\pi^2\cdot 0.033 k^2 \int_0^L d\eta\, C_n^2(\eta)\int_0^\infty d\kappa\kappa[1 - J_0(\kappa\rho)]$ $\times \kappa^{-11/3}\exp(-\kappa^2/\kappa_m^2)\cos^2\left[\dfrac{\kappa^2(L-\eta)}{2k}\right]$ $D_s(\rho) = 2.92 k^2\rho^{5/3}\int_0^L C_n^2(\eta)\,d\eta,\ (\lambda L)^{1/2} \ll \rho$

For $L \ll l_0^2/\lambda$,

$$D_s(\rho) = 0.132\pi^2 \frac{6}{5} \Gamma\left(\frac{1}{6}\right) k^2 \kappa_m^{-5/3}$$

$$\times \left[{}_1F_1\left(-\frac{5}{6}, 1, -\frac{\kappa_m^2 \rho^2}{4}\right) - 1\right] \int_0^L C_n^2(\eta)\, d\eta$$

Spherical wave, homogeneous medium

$$D_s(\rho) = 8\pi^2 k^2 \int_0^\infty d\kappa\, \kappa\, \Phi_n(\kappa) \int_0^L d\eta \left[1 - J_0\left(\frac{\rho \kappa \eta}{L}\right)\right]$$

$$\times \cos^2\left[\frac{\kappa^2 \eta (L-\eta)}{2kL}\right]$$

$$D_s(\rho) = 1.089 k^2 C_n^2 \rho^{5/3} L - 2[B_x(0) - B_x(\rho)]^\dagger$$

Spherical wave, smoothly varying medium

$$D_s(\rho) = 8\pi^2 k^2 \int_0^\infty d\kappa\, \kappa\, \Phi_n^{(0)}(\kappa) \int_0^L d\eta\, C_n^2(\eta) \left[1 - J_0\left(\frac{\rho \kappa \eta}{L}\right)\right]$$

$$\times \cos^2\left[\frac{\kappa^2 \eta (L-\eta)}{2kL}\right]$$

$$D_s(\rho) = 8\pi^2 k^2 \int_0^L d\eta\, C_n^2(\eta) \int_0^\infty d\kappa\, \kappa \left[1 - J_0\left(\frac{\rho \kappa \eta}{L}\right)\right]$$

$$\times \kappa^{-11/3} \exp(-\kappa^2/\kappa_m^2) \cos^2\left[\frac{\kappa^2 \eta (L-\eta)}{2kL}\right]$$

Beam wave, homogeneous medium

$$D_s(\rho) = 2\pi^2 \int_0^L d\eta \int_0^\infty d\kappa\, \kappa \{[J_0(i2\gamma_2 \kappa \rho_1)$$

$$+ J_0(i2\gamma_2 \kappa \rho_2) - J_0(\kappa P) - J_0(\kappa P^*)\}|H|^2$$

$$- \{1 - J_0(\kappa Q)\}H^2 - \{1 - J_0(\kappa Q^*)\}H^{*2}]\Phi_n(\kappa)^\ddagger$$

No cases worked out.

Beam wave, smoothly varying medium Let $\Phi_n(\kappa) = C_n^2(\eta)\Phi_n^{(0)}(\kappa)$ in previous formula. No cases worked out.

---

* $\Phi_n(\kappa) = 0.033\, C_n^2(x)\kappa^{-11/3} \exp(-\kappa^2/\kappa_m^2)$ except as noted. For homogeneous medium, $C_n^2(x) =$ constant.
† $\Phi_n(\kappa) = 0.033\, C_n^2 \kappa^{-11/3}$.
‡ Symbols defined in Tables 2.1 and 2.2.

CHAPTER 3

FEEDBACK IN DATA-TRANSMISSION SYSTEMS

Preview
There are various ways of introducing feedback in a data-transmission system. Two principal categories are systems with predecision feedback and systems with postdecision feedback; the latter, in turn, may be divided into information feedback systems and decision feedback systems. There are also systems with center-of-gravity information feedback.

All these systems may be divided into classes according as they have or do not have a memory; have bounded number of repetitions; are adaptive; and so on.

INTRODUCTORY REMARKS

Stringent demands are made on the noise immunity of modern data-transmission systems. These demands may be met by such effective measures as noiseproof encoding (error-detecting and error-correcting codes), various nonparametric methods, parallel-channel transmission (diversity reception), etc. One of the best-known techniques to improve noise immunity is the use of feedback (which may also be combined with other techniques; for example, a "coding-feedback" combination is often used).

In this chapter we discuss different types of feedback and the mathematical relations describing the effect of its use in a special case.

We shall assume that the system is based on transmission of discrete messages.

3.1. INTRODUCTION AND USE OF FEEDBACK — BASIC TECHNIQUES

Communication systems may be classified as one-way or two-way. The meaning of these terms is clear: in the former, information is transmitted in one direction only, say from point A to point B, whereas in the latter it may be transmitted both from A to B and from B to A. In turn, two-way systems may be either simplex (at any given instant of time, information may be transmitted either from A to B or from B to A, but not in both directions simultaneously) or duplex (transmission may be in both directions at once). It is clear that feedback may be organized only in two-way systems; a feedback path may indeed be introduced in a one-way system, but this requires special measures which considerably raise the cost of the system.

Postulating now that a feedback path is available, we turn to the question of how it is to be used. The three basic techniques are illustrated in Figure 3.1.

1. *Predecision feedback.* In this variant the feedback loop includes only the transmission line, and the information travelling along the feedback path refers in fact to the state of the forward path.

2. Feedback is introduced after the decision-making link but before the decoder. The feedback loop includes the demodulator and the signal generator, but has no influence on encoding.

3. The feedback loop encompasses the entire system, including encoder and decoder.

Types 2 and 3 are special cases of postdecision feedback.

In a postdecision feedback system, the feedback signals give instructions for retransmission of incorrectly or unreliably received signals, or for transmission of other signals representing the same information. In this connection one

Sec. 3.1. INTRODUCTION AND USE OF FEEDBACK

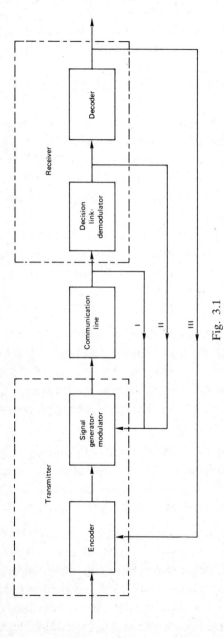

Fig. 3.1 Basic techniques for introduction of feedback in communications systems: I. Predecision feedback; II, III. postdecision feedback.

distinguishes between information feedback systems (IFS) and decision feedback systems (DFS).

In an IFS, the receiver tells the transmitter what signals it has received. The transmitter checks this information against the actually transmitted signals and, if there is a discrepancy, sends additional information about the incorrectly received elements.

In a DFS, the receiver itself, using some prescribed criterion, identifies insufficiently reliable signals and sends an interrogating signal along the feedback path. Upon reception of this signal, the transmitter again sends additional information about the unreliable elements of the message.

Thus, both IFS and DFS involve transmission of additional information about certain message elements, but just what information is needed is decided in an IFS by the transmitter and in a DFS by the receiver.

IFS, in turn, may be divided into the following variants:

1. Systems with a) complete and b) imcomplete comparison. In the first case, the entire received signal is relayed along the feedback path. In the second, the signals are divided into several groups, and the information transmitted along the feedback path specifies only the group to which the received signal belongs.

2. Systems with a) storage and b) dumping. In the first variant, the transmitter stores and later uses the signals received from the feedback path. In the second, these signals are checked and immediately erased ("dumped").

3. Systems with a) unlimited and b) limited correction. In the former, transmission is repeated until all errors have been corrected. In the latter, time limits are imposed on the transmission and repetition of each signal (or signal group); after the lapse of a certain time, the system automatically proceeds to transmission of the next signal(s) (regardless of whether all errors have been corrected).

Sec. 3.1. INTRODUCTION AND USE OF FEEDBACK

There are also different variants of DFS. In order to design a DFS or analyze its performance, the following questions must be settled:

1. Limited or unlimited interrogation? In the first case, a limited time T is assigned to the transmission of each signal (or signal group); in the second case, the system continues transmitting each signal until the receiver has received it with the desired degree of reliability. In the first case, the decision concerning a specific signal may also be made in less time than T.

2. Does the system have a memory? If it does, the decision-making process may also utilize previously received signals; otherwise only the signal forthcoming at the time in question is used.

3. Is the system adaptive? In an adaptive system, the type of signal and/or the method of reception may differ from one repetition to the next. This is of course impossible in a nonadaptive system.

4. Does interrogation refer to individual elements of the signal or to whole groups of elements?

It should thus be clear that both IFS and DFS may be implemented in quite different variants.

Yet another interesting feedback modification is center-of-gravity information feedback [35]. Here the system spends a fixed time T in transmitting each message element. The decision must always be made within time T, so that in this sense there is no difference between this and a system without feedback. However, during the time T the receiver notifies the transmitter as to the state of all its "knowledge" concerning the transmitted symbol, and the transmitter uses this information to control the transmission. The term "center-of-gravity information feedback" is motivated by the similarly between a) the equation describing the operation of the transmitter after it receives information from the feedback path, and b) the equation of classical mechanics for the center of gravity of a given system of material points.

Of the various types of feedback system discussed above, we select the DFS for more detailed analysis and mathematical presentation. We shall assume that both receiver and transmitter are provided with a memory sufficiently large to store forthcoming information for any length of time, whence it may be accessed at any desired time.

3.2. MATHEMATICAL DESCRIPTION OF FEEDBACK SYSTEMS

We shall consider a DFS with interrogation for individual message elements. Let us assume that in order to transmit M messages $\{s_j\}$, $j < M$, the transmitter uses a group of signal sequences $\{y_j^i\}$, and the receiver partitions the signal space into M acceptance regions G_j and an uncertainty region B. The transmission of a message s_j may then be described as follows. The transmitter emits a signal y_j^1; owing to channel noise, the signal is distorted, and the signal arriving at the receiver is $\tilde{y}_j^1$. If this signal is in one of the regions G_j, say G_q, the receiver decides that the transmitted signal was s_q, and notifies the transmitter (along the feedback path) that a decision has been made and the next message may be sent. But if the signal $\tilde{y}_j^1$ is in the region B, no final decision is taken, and the transmitter receives a request for additional information. Accordingly, the transmitter sends a signal y_j^2, and the signal $\tilde{y}_j^2$ at the receiver end is processed in similar fashion.

If the system is designed with limited interrogation, the region B shrinks at some stage to zero size (i.e., becomes the empty set), and the transmission process automatically comes to an end. If the set B is never empty, we have a system with unlimited interrogation.

Note that in a system with memory, decisions are based on the processing of several received signals $\tilde{y}_j^i$, whereas in a

Sec. 3.2. MATHEMATICAL DESCRIPTION OF FEEDBACK SYSTEMS

system without memory only the last of these signals is involved. In an adaptive system the sets $\{G_j\}$ may change, in a nonadaptive system they are constant.

We now describe the situation in terms of probability theory.

Let H_j be the event: message s_j is sent. Let $P(H_j|H_i)$ denote the probability that the receiver makes the decision H_j (i.e., decides that the transmitter has sent s_j) whereas in reality the signal sent was s_i.

Assuming that the system is nonadaptive, without memory and with error-free feedback path, we can write the probability that message s_j will be correctly received as

$$P(H_j|H_j) = P(\tilde{y}_j^1 \in G_j) + P(\tilde{y}_j^2 \in G_j; \tilde{y}_j^1 \in B)$$
$$+ P(\tilde{y}_j^3 \in G_j; \tilde{y}_j^1, \tilde{y}_j^2 \in B) + \cdots, \qquad (3.1)$$

where $P(H_j|H_j)$ is the probability of correct reception of s_j, $P(\tilde{y}_j^i \in G_j)$ is the probability that $\tilde{y}_j^i$ is a point of G_j, and $P(\tilde{y}_j^i \in B)$ the probability that $\tilde{y}_j^i$ is a point of B.

The assumption that the feedback path is error-free will be retained throughout the sequel. In some cases it is justified. For example, in "satellite/Earth" communication, the "Earth-to-satellite" feedback path is practically absolutely reliable, thanks to the high power of the Earth-bound transmitter. Other cases may require additional investigation; an example will be given in §4.5.

With $P(H_j|H_j)$ determined from (3.1), it is easy to determine the probability of an error in transmission of s_j:

$$\mathbf{P}_j = 1 - P(H_j|H_j), \qquad (3.2)$$

and the total error probability of the system is

$$\mathbf{P} = \sum_j P(H_j) \cdot \mathbf{P}_j, \qquad (3.3)$$

where $P(H_j)$ is the prior probability of message s_j and $\mathbf{P}_j$ is defined by (3.2).

The mean number of signals y_j^i participating in transmission of s_j is

$$K_j = 1 + \mathbf{P}(\tilde{y}_j^1 \in B) + \mathbf{P}(\tilde{y}_j^1 \in B; \tilde{y}_j^2 \in B) + \cdots. \quad (3.4)$$

These formulas are considerably simplified if one adopts the following assumptions:
1) for given j, the y_j^i are identical, $y_j^i = y_j$;
2) the events $\tilde{y}_j^i \in G_p$ and $\tilde{y}_j^q \in G_m$ are independent if $i \neq q$.

Then Eq. (3.1) becomes

$$\mathbf{P}(H_j | H_j) = \mathbf{P}(\tilde{y}_j \in G_j) + \mathbf{P}(\tilde{y}_j \in B) \cdot \mathbf{P}(\tilde{y}_j \in G_j)$$
$$+ [\mathbf{P}(\tilde{y}_j \in B)]^2 \cdot \mathbf{P}(\tilde{y}_j \in G_j) + \cdots \quad (3.5)$$
$$= \mathbf{P}(\tilde{y}_j \in G_j)[1 + \mathbf{P}(\tilde{y}_j \in B) + \mathbf{P}^2(\tilde{y}_j \in B) + \cdots].$$

If the transmission process for s_j is limited to L_j repetitions, one has to modify the G_j's at the last (L_j-th, say) step in order to make B the empty set, and

$$\mathbf{P}(H_j | H_j) = \mathbf{P}(\tilde{y}_j \in G_j) + \mathbf{P}(\tilde{y}_j \in B) \cdot \mathbf{P}(\tilde{y}_j \in G_j)$$
$$+ [\mathbf{P}(\tilde{y}_j \in B)]^2 \cdot \mathbf{P}(\tilde{y}_j \in G_j) + \cdots$$
$$+ [\mathbf{P}(\tilde{y}_j \in B)]^{L_j-2} \cdot \mathbf{P}(\tilde{y}_j \in G_j)$$
$$+ [\mathbf{P}(\tilde{y}_j \in B)]^{L_j-1} \cdot \mathbf{P}(\tilde{y}_j \in G_j^{\text{last}}), \quad (3.5a)$$

where G_j^{last} is the modified G_j of the last step. From (3.5a) we obtain

$$\mathbf{P}(H_j | H_j) = \mathbf{P}(\tilde{y}_j \in G_j) \cdot \frac{1 - \mathbf{P}^{L_j-1}(\tilde{y}_j \in B)}{1 - \mathbf{P}(\tilde{y}_j \in B)}$$
$$+ [\mathbf{P}(\tilde{y}_j \in B)]^{L_j-1} \cdot \mathbf{P}(\tilde{y}_j \in G_j^{\text{last}}). \quad (3.6)$$

If the process is unlimited, $L_j \to \infty$, and

$$\mathbf{P}(H_j | H_j) = \frac{\mathbf{P}(\tilde{y}_j \in G_j)}{1 - \mathbf{P}(\tilde{y}_j \in B)}. \quad (3.7)$$

Similarly, the mean number of repetitions is

$$K_j = 1 + \mathbf{P}(\tilde{y}_j \in B) + \mathbf{P}^2(\tilde{y}_j \in B) + \cdots, \quad (3.8)$$

Sec. 3.2. MATHEMATICAL DESCRIPTION OF FEEDBACK SYSTEMS

so that for a limited number of repetitions L_j

$$K_j = \frac{1 - \mathbf{P}^{L_j}(\tilde{y}_j \in B)}{1 - \mathbf{P}(\tilde{y}_j \in B)}, \qquad (3.9)$$

and if the number of repetitions is unlimited,

$$K_j = \frac{1}{1 - \mathbf{P}(\tilde{y}_j \in B)}. \qquad (3.10)$$

The above formulas (3.7) through (3.10) are cited from [18]; in Chapter 4 (mainly in §4.3) they will be used to evaluate the noise immunity of laser feedback systems. We shall then use the fact that if the energy of the signal y_j is W_j, then the mean energy required to transmit s_j is

$$\bar{W}_j = W_j \cdot K_j = W_j \frac{1 - \mathbf{P}^{L_j}(\tilde{y}_j \in B)}{1 - \mathbf{P}(\tilde{y}_j \in B)}. \qquad (3.11)$$

It remains only to remark that, to partition the signal space into regions G_j, the receiver may compute the likelihood ratios between pairs of signals [18], subsequently comparing these ratios with given thresholds. The thresholds themselves may be optimized [18].

CHAPTER 4

NOISE IMMUNITY OF LASER COMMUNICATION SYSTEMS AND ITS IMPROVEMENT BY USE OF FEEDBACK

Preview
Optical communication systems using time division multiplexing are more immune to noise than those using frequency division multiplexing.

The use of feedback considerably improves the noise immunity of a data system. The maximum noise immunity of a system depends on the noise immunity of a certain sequential algorithm.

In any specific system, the efficient use of feedback depends on the following factors: whether the system has a memory, fluctuations of the optical signal at the receiver, noise level, the cost of energy in the feedback path, and the maximum number of repetitions in the system.

INTRODUCTORY REMARKS

This chapter will deal with the main topic of the book: the effect of feedback on noise immunity of LCS, the advantages gained by using feedback under various conditions, the optimum mode of operating LCS with feedback, and various factors that determine whether feedback should or should not be introduced. In view of the great variety of feedback systems and types of LCS, it would be well-nigh impossible (and hardly useful) to consider all possible variants, and the only way we can demonstrate the introduction of feedback is to do so for some preselected type of LCS. Our first task will therefore be to select this basic LCS.

All LCS discussed henceforth transmit binary information with symbol-by-symbol interrogation.

The exposition is arranged as follows. In §4.1 we select the basic LCS. Then, in §4.2, we determine the maximum noise immunity of the basic LCS when feedback is used. To do this, we adopt the following assumptions: 1) The system is provided with a memory. 2) The received signals are free of fluctuations. 3) The feedback path is error-free. 4) The number of repetitions is unlimited. Each of these assumptions will be dropped in turn, in order to ascertain the significance of the relevant factor for the noise immunity and efficiency of feedback systems. Thus, in §4.3 we study memoryless systems, in §4.4 we analyze the effect of atmospheric signal fluctuations, and in §4.5 the effect of errors in the feedback path and limitations on the number of repetitions. For the most part, the chapter is an exposition of the author's own papers.

4.1. SELECTION OF BASIC LCS. NOISE IMMUNITY OF ONE-WAY SYSTEMS

The first question to be decided is whether the receiver of our hypothetical LCS should utilize optical heterodyning. On the one hand, if there is a high background noise level, optical heterodyning provides a marked improvement in noise immunity. On the other, use of the principle introduces two complex problems: a) matching of the signal wave fronts and the local oscillator; b) mutual stabilization of signal and oscillator as to frequency and phase. Moreover, both these problems must be solved to within a precision compatible with the optical wavelength and with the period of the optical carrier, respectively. The attendant technical difficulties usually discourage the designer from using heterodyning, all the more so as modern optical filters considerably lower the background noise level at the detector stage and thus nullify the advantages of heterodyning.

For these reasons, although some circumstances do justify optical heterodyning, we shall henceforth avoid its use in LCS. At the same time, we expect our conclusions to be qualitatively valid for LCS with heterodyning as well.

Sec. 4.1. SELECTION OF BASIC LCS

The next problem is the type of multiplexing to use. LCS are usually intended for transmission of large amounts of information over several independent channels. In principle, these channels may be multiplexed in two ways: a) use of different optical carriers; b) use of a single optical carrier with time division or subcarrier frequency division multiplexing (see below). In the first case, transmission is accomplished by simply combining several laser beams (depending on the number of channels). Though attractively simple (at least, in theory), this approach ignores the question of whether each individual optical carrier can be multiplexed and does not use the carrier capacity to the full. Moreover, when the number of channels is high the method is uneconomical (unless several channels are first multiplexed on one carrier).

We therefore proceed to consider multiplexing of a single optical carrier (of course, several multiplexed carriers may be utilized together, with frequency diversity, space diversity, etc.)

When a single optical carrier is being used, with intensity-modulated radiation, one may apply either time division or frequency division multiplexing with subcarrier waves. As PCM (pulse-code modulation), one of the most promising systems in use today, is essentially equivalent to transmission of binary signals, we confine attention from the start to this case (Figure 4.1).

The function $\bar{\lambda}(t)$ shown in the figure represents the first moment function of the detector photocurrent (which is proportional to the intensity of the optical signal); the digits 0 and 1 represent the transmitted signals.

In frequency division multiplexing, each subcarrier wave (only one is shown in the figure) may be modulated with passive spacing[8] (amplitude modulation — AM) or with

---

[8] By "passive spacing" we mean a system of modulation that uses the signal $S_0 = 0$ to transmit the message element 0 and a signal S_1, different

68 Ch. 4. NOISE IMMUNITY OF LCS

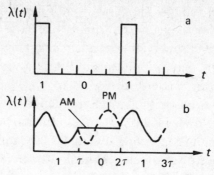

Fig. 4.1
Mean photocurrent of signal for different multiplexing methods:
a) time multiplexing (intensity modulation), b) frequency multiplexing (AM — subcarrier amplitude modulation; PM — subcarrier phase modulation). (Signal of only one channel shown.)

active spacing (phase or frequency modulation — PM or FM). In the case of time division multiplexing we shall limit ourselves for the moment to passive spacing (active spacing will be considered in the next section).

To select the type of multiplexing, let us determine the potential noise immunity of LCS using the different systems, and select the better of the two.

We first consider a general problem: optimum reception of discrete optical signals, the criterion being the likelihood function for a binary channel (both cases reduce to this problem). Let us assume that the signals (elementary pulses) are identified by their first moment functions $\bar{\lambda}_k(t)$ ($k = 0, 1$), measured by the number of detector photoelectrons per second (the subscript k indicates the specific message element). The function $\bar{\lambda}_k(t)$ depends both on the signal and on the background noise.

At this stage, we assume that the optical signal does not fluctuate (fluctuating signals will be considered in §4.4), and

---

from S_0, to transmit the element 1. "Active spacing" also employs two different signals S_0 and S_1, but each of them must differ from zero at least at one point.

that the detector is tuned to photon counting (the choice of detector may be based on considerations and calculations as in §1.3).

In §2.1 we discussed some cases in which (under certain restrictions) the photocurrent is Poisson-distributed; at least one of these cases is characteristic for LCS: single-mode coherent signal, multimode noise with small occupation number per mode. In view of this and other similar cases, we shall treat the photocurrent as a Poisson stream of electrons, generally nonstationary (because of modulation);[9] as such, it is fully characterized by its mean $\bar{\lambda}_k(t)$.

Let us divide the signal duration into elementary intervals Δt_i and consider $\bar{\lambda}(t) = \bar{\lambda}(t_i)$ as constant within each interval; for each Δt_i, we determine the elementary likelihood ratio

$$\Lambda_i = \frac{\mathbf{P}[1|\lambda(t_i)]}{\mathbf{P}[0|\lambda(t_i)]}, \qquad (4.1)$$

where $\mathbf{P}[k|\lambda_i]$ is the posterior probability of signal k if the detector current at time t_i is $\lambda(t_i)$. Then, multiplying the Λ_i together, letting $\Delta t \to 0$ and taking logarithms, we obtain

$$\ln \Lambda = \ln \frac{\mathbf{P}(1|\lambda)}{\mathbf{P}(0|\lambda)} \cdot z + k, \qquad (4.2)$$

where

$$z = \int_0^\tau \lambda(t) l(t) dt; \qquad l(t) = \ln \frac{\bar{\lambda}_1(t)}{\bar{\lambda}_0(t)};$$

$$k = \int_0^\tau [\bar{\lambda}_0(t) - \bar{\lambda}_1(t)] dt + \ln \frac{\mathbf{P}(1)}{\mathbf{P}(0)};$$

$\mathbf{P}(k)$ $(k = 0, 1)$ are the prior probabilities of the two alternative hypotheses (types of message element), and $\mathbf{P}(k|\lambda)$ $(k = 0, 1)$ are their posterior probabilities given the sampled signal $\lambda(t)$ over the signal interval τ. Note that z is a random variable, and k is a constant which vanishes when the

---

[9] This assumption will be adhered to throughout the chapter.

hypotheses have equal prior probabilities (as is typical in communication systems) and the optical signals have equal mean energies (as in subcarrier systems).

Equation (4.2) provides us with an algorithm for optimum reception; it remains only to determine the corresponding error probability.

In the special case of detection of square pulses (Figure 4.1a) $l(t) = l = $ const, and the calculations, including determination of the optimum threshold and error probability, involve no difficulties, as the variable z/l is Poisson-distributed.

In general, however, with $l(t)$ an arbitrary function, derivation of an exact formula for the distribution of z is an extremely difficult task. Attempts to approximate the distribution of the function $\lambda(t)$ by a normal law produce a highly inferior optimum reception algorithm and the resulting estimates of noise immunity are very coarse.

Nevertheless, if we retain the optimum reception algorithm for a frequency division multiplex system as

$$z = \int_0^T \lambda(t)l(t)dt \begin{cases} > 0 \to 1, \\ < 0 \to 0 \end{cases} \qquad (4.3)$$

(as follows from (4.2)), assuming as before that the incident beam is a nonstationary Poisson stream of electrons, it follows from the central limit theorem that the distribution of z should be "nearly" normal. On this assumption, let us express the integral distribution function of z as a Gram-Charlier series [40]:

$$F_z = F\left(\frac{z - m_1}{\sigma}\right)$$

$$-\frac{1}{\sqrt{2\pi}} \exp\left[-\frac{(z - m_1)^2}{2\sigma^2}\right] \sum_{k=3}^{\infty} \frac{c_k}{k!} H_{k-1}\left(\frac{z - m_1}{\sigma}\right), \qquad (4.4)$$

where $H_k(x)$ are the Hermite polynomials, $c_k = \int_{-\infty}^{\infty} H_k(z)w_z\,dz$; w_z, m_1, σ^2 are the probability density, expectation and variance of z, respectively;

$$F(y) = \frac{1}{\sqrt{2\pi}} \int_{-\infty}^{y} e^{-x^2/2} dx.$$

Our next task is to find the moments of the random variable x, which we need for the expansion. The characteristic function of z is

$$\theta_z = \exp \int_0^T \bar{\lambda}(t)(e^{i\omega l(t)} - 1) dt. \qquad (4.5)$$

The cumulant function of z and its derivatives with respect to ω are

$$\psi_z = \int_0^T \bar{\lambda}(t)(e^{i\omega l(t)} - 1) dt \quad \text{and} \quad \psi_z^{(s)} = i^s \int_0^T l^s(t) \lambda(t) e^{i\omega l(t)} dt. \qquad (4.6)$$

The first four moments of z, the skewness coefficient α and the coefficient of excess γ are

$$m_1 = \frac{1}{i} \theta_z^{(1)}(0) = \int_0^T \bar{\lambda} l \, dt; \quad M_2 = \sigma^2 = -\psi_z^{(2)}(0)$$

$$= \int_0^T \bar{\lambda} l^2 dt;$$

$$M_3 = i\psi_z^{(3)}(0) = \int_0^T \bar{\lambda} l^3 dt; \quad M_4 = \psi_z^{(4)}(0) + 3M_2^2$$

$$= \int_0^T \bar{\lambda} l^4 dt + 3M_2^2; \qquad (4.7)$$

$$\alpha = M_3 M_2^{-3/2}; \quad \gamma = M_4 M_2^{-2} - 3.$$

Limiting ourselves to the first three terms of the Gram–Charlier series (4.4), we obtain

$$F_z = F\left(\frac{z - m_1}{\sigma}\right) - \frac{1}{\sqrt{2\pi}} e^{-(z - m_1)^2 / 2\sigma^2} \left[\frac{\alpha}{3!} H_2\left(\frac{z - m_1}{\sigma}\right) + \frac{\gamma}{4!} H_3\left(\frac{z - m_1}{\sigma}\right) \right], \qquad (4.8)$$

where $H_2(x) = x^2 - 1$, $H_3(x) = x^3 - 3x$.

Ch. 4. NOISE IMMUNITY OF LCS

The conditional error probabilities may now be found from the formulas

$$P(1|0) = F_{z_0}(0) \quad \text{and} \quad P(0|1) = 1 - F_{z_1}(0), \quad (4.9)$$

where the subscripts 0 and 1 refer to the two elementary pulses $\bar{\lambda}_0(t)$ and $\bar{\lambda}_1(t)$.

When subcarrier waves are used, in one channel:

for signal "0": $\bar{\lambda}_0(t) = \bar{\lambda}_s(1 + m_0 \sin \Omega t) + \bar{\lambda}_N$,

for signal "1": $\bar{\lambda}_1(t) = \bar{\lambda}_s(1 + m_1 \sin \Omega t) + \bar{\lambda}_N$, (4.10)

where $\bar{\lambda}_s$ is the mean photocurrent induced by the signal, $\bar{\lambda}_N$ is the mean photocurrent due to background noise, Ω and m are the frequency and depth of modulation ($m_0 = 0$ in AM, $m_0 = -m_1$ in PM).

We must now substitute (4.10) into (4.7), (4.7) into (4.8), and (4.8) into (4.9); this was the method used in [39] to calculate the error probability on the assumption that the prior probabilities of the elements are equal:

$$P(0) = P(1) = \frac{1}{2}; \quad P = \frac{1}{2}[P(1|0) + P(0|1)]. \quad (4.11)$$

The results of a computer calculation are shown in Figure 4.2.

The effect of background noise uncorrelated with the signal (multimode fog, dark current) is to reduce the effective depth of modulation by a factor of $1 + \bar{\lambda}_N/\bar{\lambda}_s$:

$$\bar{\lambda}(t) = \bar{\lambda}_s(1 + m \sin \Omega t) + \bar{\lambda}_N = (\bar{\lambda}_s + \bar{\lambda}_N)\left(1 + \frac{m \sin \Omega t}{1 + \bar{\lambda}_N/\bar{\lambda}_s}\right).$$
(4.12)

For this reason, the parameter of the curves in Figure 4.2 is the effective depth of modulation:

$$m_e = m \cdot \frac{1}{1 + \bar{\lambda}_N/\bar{\lambda}_s}. \quad (4.13)$$

Sec. 4.1. SELECTION OF BASIC LCS 73

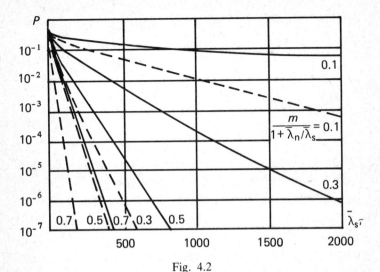

Fig. 4.2
Noise immunity of LCS with frequency multiplexing. Solid lines show amplitude modulation, dashed lines — phase modulation.

We can now check whether the distribution (4.4) can be approximated by a normal law. To this end, we repeat the error-probability computation, using only the first term of (4.4). The difference between the computed values of p appear only in the third significant digit, so that for engineering purposes it is quite legitimate to assume that z is normally distributed.

We can now utilize the above results to select the type of multiplexing most immune to noise. To do this, we compare the signal powers that yield the same error probability, given the number q of channels.

Let $\bar{\lambda}_N = 0$, $q = 3$, $p = 10^{-6}$. Considering an LCS with depth of modulation $m = 1/3$, which is reasonable in regard to minimization of nonlinear distortions and interchannel interference, we see from Figure 4.2 that $\bar{\lambda}_s \tau = 1700$ for AM, $\bar{\lambda}_s \tau = 460$ for PM.

On the other hand, in time division multiplexing the optimum algorithm reduces to summation and, as is readily seen, the error probabilities are defined by

$$\alpha = \mathbf{P}(1|0) = \sum_{i=d+1}^{\infty} \frac{(\bar{n}_N)^i}{i!} e^{-\bar{n}_N} = 1 - \sum_{i=0}^{d} \frac{(n_N)^i}{i!} e^{-\bar{n}_N},$$

$$\beta = \mathbf{P}(0|1) = \sum_{i=0}^{d} \frac{(\bar{n}_1)^i}{i!} e^{-\bar{n}_1}, \qquad (4.14)$$

$$\mathbf{P} = \alpha \cdot \mathbf{P}(0) + \beta \cdot \mathbf{P}(1)$$

(if $\mathbf{P}(0) = \mathbf{P}(1) = \frac{1}{2}$, then $p = \frac{1}{2}(\alpha + \beta)$), where $\bar{n}_s = \bar{\lambda}_s \tau$ is the signal energy for each elementary pulse, $\bar{n}_N = \bar{\lambda}_N \tau$ is the background noise energy for each elementary pulse, $\bar{n}_1 = \bar{n}_s + \bar{n}_N$, and $d = \bar{n}_s / \ln(S+1)$ is the threshold, rounded off to the least integer.

Using Eqs. (4.14), one readily sees that to guarantee the same error probability in a system with time division multiplexing one needs only 14 photoelectrons per pulse. In a three-channel system with the same data rate, the mean number of photoelectrons[10] = number of channels × number of photoelectrons per elementary pulse × probability of active elementary pulse = $14 \times 3 \times \frac{1}{2} = 21$.

In other words, frequency division multiplexing is energetically inferior to time division multiplexing: the energy loss is approximately 20 db in AM, 15 db in PM. And this is true regardless of our assumption that the phase of the subcarrier waves is known (phase indeterminacy leads to a further deterioration of noise immunity [41]).

In accordance with our maximum noise immunity criterion, therefore, we decide to use systems with time division multiplexing.

To summarize: our hypothetical LCS will employ time division multiplexing, and its receiver will not be a heterodyne device.

Our next goal is to determine the rise in noise immunity of the LCS achieved by using various types of feedback.

---

[10] This calculation includes only the active transmission, since the transmitter consumes no energy in passive transmission.

4.2. MAXIMUM NOISE IMMUNITY OF FEEDBACK SYSTEMS. SYSTEMS WITH MEMORY. SEQUENTIAL ANALYSIS

We begin our analysis of feedback systems with a calculation of their maximum noise immunity. To this end, we shall utilize the general formula for noise immunity of a feedback system and optimize it as a function of the system parameters.

This rather complex problem is considerably simplified if we take the following considerations into account:

1. The procedure whereby the receiver makes a decision with regard to a message a) involves testing of statistical hypotheses, and b) is implemented "step by step,"[11] i.e., is a sequential procedure.

2. Of all sequential (and nonsequential) tests of equal strength, for statistical hypotheses, the Wald sequential procedure requires minimum mean sample size.

Points 1 and 2 in themselves are sufficient to imply that the noise immunity of a feedback system cannot exceed that of a sequential algorithm; we now show that it can in fact approach this limiting value.

3. We consider a special type of feedback system — a system with interrogation. The receiver divides the signal space into acceptance regions G_k and an indeterminate region B (or null region). If the signal belongs to region G_k, the decision taken is H_k (k-th signal transmitted) and the transmitter is notified (over the feedback path) that a decision has been made and transmission of the element in question may be ended. If the signal is found to lie in the null region, no final decision is made and the transmitter is

---

[11] Details on the operating principle and physical realization of feedback systems may be found in Chapter 3 or (briefly) in §4.3.

requested to relay additional information about the message element.

4. A system with interrogation may be provided with a memory. In that case the receiver stores the results of *all* decisions concerning elementary pulses relating to the message element, and can take them into account in future decisions.

5. If we assume that the feedback path is error-free and the number of repetitions unlimited, the decision-making algorithm is an untruncated Wald sequential algorithm with grouping [17]. To improve noise immunity, we must eliminate grouping, i.e., make the energy per elementary pulse tend to zero. The noise immunity will then tend to that of an untruncated Wald algorithm without grouping, which we henceforth refer to in brief as a sequential algorithm.

In other words, in order to determine the optimum noise immunity of our LCS with feedback, it will suffice to do this for the appropriate sequential algorithm.

The analysis will be carried out for both passive and active spacing.

Let us consider a sequential algorithm for the previous case — coherent single-mode signal on the background of additive multimode noise with small occupation number per mode. We already know that the number of photons (and in case of direct detection also the number of photoelectrons in the detector) is Poisson-distributed, whether there is or is not a signal; the expectation and fluctuations (quantum noise) depend both on the signal and on the additive background noise.

It should be clear that these relationships are valid for other cases characterized by a Poisson distribution (e.g., detection of a single, incoherent signal with small occupation numbers on the background of another signal of the same type, and so on).

From the mathematical standpoint, then, we have to

Sec. 4.2. MAXIMUM NOISE IMMUNITY OF FEEDBACK SYSTEMS

determine the characteristics of a sequential test for hypotheses about the parameter of a Poisson distribution — this problem is familiar in the case of the normal and binomial distributions [17, 42].

In terms of communication theory, this is equivalent to finding the maximum noise immunity of an LCS with interrogation; naturally, the maximum is unattainable in a real LCS with interrogation and memory, but it may nevertheless be approached arbitrarily closely. We shall compare the results with the situation for a deterministic sample (one-way system), in regard to the mean energy requirements for a prescribed error probability. The savings in mean energy will indicate the efficiency of the sequential analysis (system with interrogation) for transmitters of limited mean power (in the optical region, these are semiconductor lasers).

A. Passive spacing

The system uses amplitude modulation. In case of binary signals, the receiver must solve a detection problem, adopting one of two decisions:

a) the channel contains only background noise;
b) the channel contains both signal and background noise.

The analysis will be carried out for receivers with direct detection; the received signal (detector photocurrent) is represented by a Poisson sequence of quantum transitions.

A.1. *Formulation of the problem.* The Poisson distribution is uniquely determined by its expectation [43]. In mathematical terms, therefore, the problem of distinguishing between two nonstationary Poisson-distributed signals, which includes the detection problem as a special case, is as follows.

We have to test the hypothesis H_1 that the first moment function of the photocurrent is

$$\bar{\lambda}(t) = \bar{\lambda}_1(t)$$

against a single alternative H_0:

$$\bar{\lambda}(t) = \bar{\lambda}_0(t).$$

Then, if $\bar{\lambda}(t)$ is expressed in photoelectrons per second, the logarithm of the likelihood ratio is given by Eq. (4.2), which we write as

$$\ln \Lambda = \sum_{i=1}^{N} \lambda(t_i) \Delta t \ln \frac{\bar{\lambda}_1(t_i)}{\bar{\lambda}_0(t_i)} - \sum_{i=1}^{N} [\bar{\lambda}_1(t_i) - \bar{\lambda}_0(t_i)] \Delta t$$

$$\ln \frac{P(H_1)}{P(H_0)} = \sum_{i=1}^{N} z_i + \ln \frac{P(H_1)}{P(H_0)},$$

or

$$\ln \Lambda = \int_0^\tau \lambda(t) \ln \frac{\bar{\lambda}_1(t)}{\bar{\lambda}_0(t)} dt - \int_0^\tau [\bar{\lambda}_1(t) - \bar{\lambda}_0(t)] dt + \ln \frac{p_1}{p_0},$$
(4.15)

where N is the number of samples (in the discrete case), $\lambda(t)$ is the received photocurrent sample (photoelectrons per second), τ its duration, $\bar{\lambda}_k(t)$ its first moment function if hypothesis H_k is true ($k = 0, 1$), and $P(H_k)$ is the prior probability of H_k.

The Wald sequential algorithm involves comparing the likelihood ratio with two thresholds A and B, which are expressed in terms of α and β — the maximum probabilities of a "false alarm" $P(H_1|H_0)$ and of a "lost" pulse $P(H_0|H_1)$ [17]:

$$A = (1 - \beta)/\alpha; \quad B = \beta/(1 - \alpha). \quad (4.16)$$

Thus, the sequential algorithm divides the signal space into acceptance regions G_k ($k = 0, 1$) and an indeterminate or null region C:

$$G_1: \Lambda > A \to H_1;$$
$$G_0: \Lambda < B \to H_0; \quad (4.17)$$
$$C: B < \Lambda < A \to t \uparrow.$$

Sec. 4.2. MAXIMUM NOISE IMMUNITY OF FEEDBACK SYSTEMS

Now consider the detection of a steady signal on the background of steady noise: this means that by assumption $\bar{\lambda}_k(t) \equiv \text{const} \equiv \bar{\lambda}_k$ is a rectangular envelope. We also assume that the prior probabilities of the hypotheses are equal:

$$P(H_1) = P(H_0) = \tfrac{1}{2}.$$

Then Eq. (4.15) is simplified (its physical implementation involves only integration), and Eq. (4.17) may be written as

$$G_1: n = \int_0^\tau \lambda(t)dt \geq \frac{\bar{\lambda}_s \tau + \ln A}{\ln(S+1)} = A' \to H_1;$$

$$G_0: n = \int_0^\tau \lambda(t)dt \leq \frac{\bar{\lambda}_s \tau + \ln B}{\ln(S+1)} = B' \to H_0; \quad (4.18)$$

$$C: B' < n = \int_0^\tau \lambda(t)dt < A' \to \tau_1$$

where $\bar{\lambda}_s = \bar{\lambda}_1 - \bar{\lambda}_0$ is the mean signal photocurrent (photoelectrons per second), $\bar{\lambda}_0$ the mean photocurrent due to additive background noise (photoelectrons per second), $S = \bar{\lambda}_s/\bar{\lambda}_0$ is the signal/noise ratio.

We now determine the characteristics of the untruncated sequential algorithm (4.18): the mean analysis time for both hypotheses, the mean signal energy and the distribution function of analysis time; it will be assumed that α, β and S are known.

A.2. Characteristics of sequential detection algorithm in Wald approximation. We wish to determine the characteristics of algorithm (4.18) in the Wald approximation, i.e., on the assumption that at the time of decision the cumulative sum n coincides with A' and/or B'.

Wald [17] adopts a similar assumption ("neglecting excess" over the thresholds), obtaining the following expressions for the expected sample size:

$$E_0(N) = \frac{(1-\alpha)\ln B + \alpha \ln A}{E_0(z_i)},$$

$$E_1(N) = \frac{\beta \ln B + (1-\beta)\ln A}{E_1(z_i)}, \quad (4.19)$$

where $\mathbf{E}_k$ is the expectation symbol (the subscript refers to the hypothesis).

Now, using Eq. (4.15), we determine $\mathbf{E}_k(z_i)$ for the Poisson distribution:

$$\mathbf{E}_k(z_i) = \bar{\lambda}_k \cdot \Delta t \cdot \ln(S+1) - \bar{\lambda}_s \cdot \Delta t. \qquad (4.20)$$

We now substitute (4.20) into (4.19) and multiply by Δt, to obtain the expected analysis time:

$$\bar{\tau}_0^{sq} = \frac{1}{\bar{\lambda}_s} \cdot \frac{\alpha \ln A + (1-\alpha)\ln B}{1 + S^{-1}\ln(S+1)}, \qquad (4.21)$$

$$\bar{\tau}_1^{sq} = \frac{1}{\bar{\lambda}_s} \cdot \frac{(1-\beta)\ln A + \beta \ln B}{(1+S^{-1})\ln(S+1) - 1}. \qquad (4.22)$$

Note that when $\alpha = \beta$ the expected analysis time is not the same for both hypotheses. This is a result of the skewness of the Poisson distribution, and will be true for any channel distribution with nonzero skewness coefficient.

There is a simple geometrical interpretation of Eqs. (4.21) and (4.22). When $\alpha, \beta \ll 1$, they define the mean analysis times for different hypotheses as the points of intersections of the threshold lines with the appropriate expectations of the samples (Figure 4.3): the straight lines A' and $\bar{\lambda}_1 \tau$ for

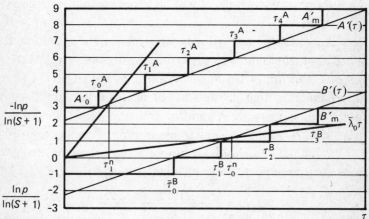

Fig. 4.3
Sequential analysis thresholds and statistical averages of random samples (passive spacing, intensity modulation).

hypothesis H_1, B' and $\bar{\lambda}_0 \tau$ for H_0. This is evident both from physical considerations and from comparison of the equations.

Using Eq. (4.22), we can determine the mean signal energy for our sequential reception algorithm:

$$\bar{n}_s^{sq} = \bar{\lambda}_s \cdot \bar{\tau}_1^{sq} = \frac{(1-\beta)\ln A + \beta \ln B}{(1 + S^{-1})\ln(S+1) - 1}. \tag{4.23}$$

This equation is considerably simplified when $\alpha = \beta = p$ in two limiting cases; using Eq. (4.15), we obtain:

$$\text{if } S \gg 1: \bar{n}_s^{sq} = \frac{-\ln p}{\ln S - 1} \simeq \frac{-\ln p}{\ln S} \quad (\text{when } \ln S \gg 1);$$
$$\tag{4.24}$$

$$\text{if } S \ll 1: \bar{n}_s^{sq} = \frac{-\ln p}{S}. \tag{4.25}$$

These noise immunity curves ($\alpha = \beta = p$) are shown in Figure 4.4. For comparison, the figure also shows the noise immunity curves in case of a deterministic sample (ordinary optimum receiver with integration), computed from Eqs. (4.14).

Using the data of Figure 4.4, we can determine $V = \bar{n}_s / \bar{n}_s^{sq}$ for fixed p — the energy efficiency of sequential analysis for detecting signals on the background of quantum noise. Some typical curves $V(S, p)$ are plotted in Figure 4.5.

The asymptotic behavior of V at p^{-1}, $S \gg 1$ is readily found in analytical form from Eqs. (4.14) and (4.24):

$$V \simeq \left(1 + \frac{\ln 2}{\ln p}\right) \ln S \simeq \ln S. \tag{4.26}$$

We now determine the distribution function of the analysis time. We introduce the notation (see Figure 4.3): τ_0^A — first time after $\tau = 0$ at which $A'(\tau)$ takes on an integer value (henceforth denoted by A_0'); τ_m^A — time at which $A'(\tau)$ takes the value $A_0' + m$ ($m = 0, 1, 2, \cdots$); τ_m^B — time at which $B'(\tau)$ takes the value m ($m = 0, 1, 2, \cdots$).

82 Ch. 4. NOISE IMMUNITY OF LCS

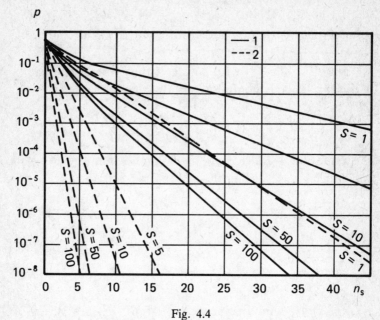

Fig. 4.4
Noise immunity of signal detection algorithms in optical channel:
1) optimal nonsequential algorithm, 2) sequential analysis.

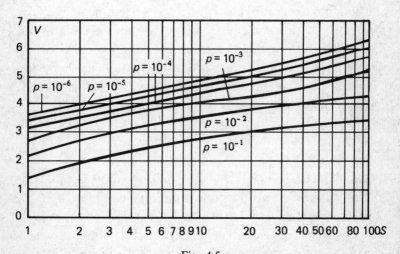

Fig. 4.5
Efficiency of sequential analysis for detecting signal in optical channels.

Sec. 4.2. MAXIMUM NOISE IMMUNITY OF FEEDBACK SYSTEMS

Note that as the photocurrent is nonnegative decision H_0 may be accepted only at times τ_m^B ($m = 0, 1, 2, \cdots$). Thus the probability $P_0(\tau_m^B)$ that decision H_0 will be made at time τ_m^B may be found by calculating the probability that m photoelectrons will be emitted by the time τ_m^B, given that the number of photoelectrons registered at each τ_l^B ($l < m$) was at least $l + 1$ (we are assuming that $P(H_0|H_0) \gg P(H_1|H_0)$, as is usually the case).

On the other hand, decision H_1 cannot be made at times τ_m^A ($m = 0, 1, 2, \cdots$). Indeed, if the decision H_1 has not been made before time $\tau_m^A - 0$, the cumulative sum at that time satisfies the inequality $n < A_0' + m$; consequently, for a decision H_1 to be accepted before time $\tau_m^A + 0$ the number of photoelectrons emitted in the time interval from $\tau_m^A - 0$ to $\tau_m^A + 0$ must be at least two; this is impossible in view of the fundamental property of the Poisson distribution: $P(n = 2) \ll P(n = 1)$ if the mean of the distribution is small [43]. Thus the cumulative distribution function $F_1(\tau)$ of decision H_1 at time τ may be found by computing the probability $1 - F_1(\tau)$ that up to time $\tau \in (\tau_m^A, \tau_{m+1}^A)$ the photoelectron count will be at most $A_0' + m$, given that at any time $\tau \in (\tau_l^A, \tau_{l+1}^A)$ ($l < m$) the count was at most $A_0' + l$ (on the assumption that $P(H_1|H_1) \gg P(H_0|H_1)$ — this condition is also usually fulfilled).

If the conditions $P(H_0|H_0) \gg P(H_1|H_0)$, $P(H_1|H_1) \gg P(H_0|H_1)$ fail to hold, the risk of wrong decisions must be taken into consideration in the probability calculations.

Thus, the distribution function of the analysis time may be determined by simply inspecting all possible alternatives with the help of the Poisson law. Analytical expressions may be found if $p^{-1} \leq S + 1 \leq p^{-2}$:

$$P_0(\tau_m^B) = e^{-\bar{\lambda}_k \tau_m^B} \cdot [\bar{\lambda}_k (\tau_0^B - \tau_0^A)]^m, \qquad (4.27)$$

$$W_1(\tau) = e^{-\bar{\lambda}_k \tau} [\bar{\lambda}_k (\tau_0^B - \tau_0^A)]^m \cdot \bar{\lambda}_k^2 \cdot (\tau - \tau_m^A): \qquad \tau_m^A \leq \tau \leq \tau_m^B,$$

$$W_1(\tau) = e^{-\bar{\lambda}_k \tau} [\bar{\lambda}_k (\tau_0^B - \tau_0^A)]^{m+1} \cdot \bar{\lambda}_k: \qquad \tau_m^B \leq \tau \leq \tau_{m+1}^A,$$

Ch. 4. NOISE IMMUNITY OF LCS

where $W_1(\tau) = \partial F_1(\tau)/\partial \tau$ is the probability density of the time at which decision H_1 is accepted.

These formulas were derived on the assumption that either of the two decisions H_1 or H_0 may be accepted; they are readily proved by induction.

The function $W_1(\tau)$ is shown in Figure 4.6, compared with the Wald distribution function [44]. It is clear that the latter provides only an approximation to the analysis-time distribution. As S decreases (and hence $\bar{n}_s^{sq}$ increases), the quality of the approximation improves thanks to the universality of the Wald distribution [44]. The case of large S will be considered in detail in the next section.

To summarize: on the assumption that at the decision-making time n coincides with A' and/or B', one can determine all the important characteristics of the algorithm; the conclusions are as follows: 1) the efficiency of sequential analysis in our context increases with increasing signal/noise

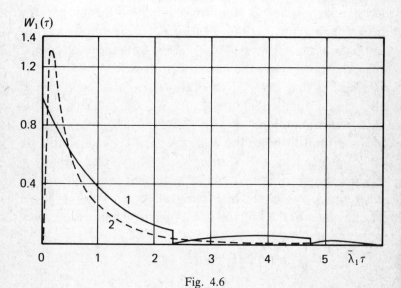

Fig. 4.6
Distribution function of analysis time for system with passive spacing
($p = 10^{-2}$, $S = 10^3$):
1) computed from exact formulas, 2) Wald distribution.

ratio S in the channel and with decreasing error probability p (Figure 4.5); 2) when $S \gg 1$ the efficiency depends only on S (see Eq. (4.26)); 3) the Wald distribution is only an approximation to the analysis-time distribution (Figure 4.6).

Strictly speaking, however, this assumption is false: n is a discrete variable, assuming values from a discrete set $(0, 1, 2, \cdots)$, whereas A' and B' are continuous. To make the results more accurate, therefore, additional analysis is necessary.

A.3. *Rigorous analysis of characteristics of sequential detection algorithm.* As hinted above, at the time of decision making the cumulative sum n coincides not with the linear functions A' and/or B', but with certain step functions A'_k and/or B'_k, obtained by rounding off A' to the next-higher integer and B' to the next-lower integer if $B' \geq 0$ (if $B' < 0$ we put $B'_k = -1$). This means that decision H_0 cannot be made before time τ_0^B (see Figure 4.1).

The thresholds A'_k and B'_k are defined by

$$A'_k = A'_0 + k; \qquad B'_k = k - 1 \qquad (k \geq 0). \qquad (4.28)$$

Our assertion that n is equal to A'_k and/or B'_k at the decision-making time is justified, if the accepted decision is H_0, by the fact that the photocurrent is nonnegative (so that this decision may be adopted only at times τ_m^B), and if the decision is H_1, by the previously mentioned property of the Poisson distribution ($\mathbf{P}(n=2) \ll \mathbf{P}(n=1)$ if the mean is small, and simultaneous emission of at least two photoelectrons is required to exceed the upper threshold).

Now, using A'_k and B'_k instead of A' and B' and setting up discrete analogs of Eqs. (4.21), (4.22), (4.23), we obtain improved values of $\bar{\tau}_0^{sq}$, $\bar{\tau}_1^{sq}$ and $\bar{n}_s^{sq}$.

In addition, if $\alpha, \beta \ll 1$, we can utilize the geometrical interpretation of Eqs. (4.21) and (4.22). In that case, the whole procedure for $\bar{\tau}_1^{sq}$ (and hence also for $\bar{n}_s^{sq}$) reduces to rounding off $\bar{\lambda}_1 \bar{\tau}_1^{sq}$ to the next-higher integer. Clearly, if our previous, nonrigorous analysis gave

$$\bar{\lambda}_1 \cdot \bar{\tau}_1^{sq} < 1, \qquad (4.29)$$

the new value will be $\bar{\lambda}_1 \bar{\tau}_1^{sq} = 1$ regardless of the actual values of α, β, S.

Using (4.16), (4.18) and (4.22), we can write condition (4.29) as

$$\frac{1-\beta}{\alpha} \leq (S+1)\exp(1+S^{-1})^{-1}. \qquad (4.30)$$

For small α, the values of S satisfying (4.30) are much larger than 1 and inequality (4.30) is simplified:

$$S \geq e \cdot \frac{1-\beta}{\alpha} \simeq \frac{e}{\alpha} = S_0. \qquad (4.31)$$

Then $\bar{\lambda}_1 \simeq \bar{\lambda}_s$ and $\bar{n}_s^{sq} = (1+S^{-1})\bar{\lambda}_1\bar{\tau}_1^{sq} \simeq \bar{\lambda}_1\bar{\tau}_1^{sq} = 1$. In other words, one quantum mean signal energy will suffice for an arbitrary small error probability (i.e., arbitrarily large reliability), provided the noise level is low enough to ensure the validity of condition (4.31); this result is in full accordance with the conclusions of information theory.

If S is sufficiently large (zero threshold d), Eq. (4.14) yields $\bar{n}_s = -\ln p$; hence it is clear that the efficiency of sequential analysis when the background noise level is asymptotically approaching zero will be higher, the higher the desired noise immunity:

$$\lim_{S\to\infty} V = -\ln p. \qquad (4.32)$$

The reader should note that this conclusion follows exclusively from our more rigorous analysis, assuming step-function thresholds; by the findings of subsection A.2, $\lim V = \ln S$ (see Eq. (4.26)).

The discrepancy between the conclusions from the approximate (A.2) and rigorous (A.3) analysis when condition (4.31) holds is an indication that the device of neglecting excess over the threshold, though frequently employed

Sec. 4.2. MAXIMUM NOISE IMMUNITY OF FEEDBACK SYSTEMS

[17, 42], is not always legitimate. Though justified in cases of high background noise, use of this assumption when the noise level is low may lead to significant errors in estimating the characteristics of the sequential algorithm.

It is a simple matter to transform this qualitative conclusion into quantitative terms: when $S \leq 0.1 S_0$, we can use the approximation (4.21)–(4.23); the error thus introduced in $\bar{n}_s^{sq}$ is at most 0.4 db. When $0.1 S_0 < S < S_0$, Eqs. (4.21)–(4.23) yield an error of up to 3 db; when $S > S_0$ we have $\bar{n}_s^{sq} = 1$, and application of Eqs. (4.21)–(4.23) involves an error that tends to infinity as $S \to \infty$.

The sequential algorithm has an interesting property under conditions of quantum noise: the algorithm for detection of a signal on a background of low-level noise is finite. Indeed, let $\tau_0^B \leq \tau_0^A$; in view of (4.16) and (4.18), this condition may be written

$$S + 1 \geq \frac{(1-\beta)(1-\alpha)}{\alpha\beta} \simeq \frac{1}{\alpha\beta}. \tag{4.33}$$

If $\beta \leq e^{-1}$ (as is usually the case), condition (4.33) is weaker than (4.31), and implies that $\bar{n}_s^{sq} = 1$. If no photoelectron has been registered in the time $\tau < \tau_0^B$ (appearance of a photoelectron in this period would imply acceptance of decision H_1), then decision H_0 is accepted at time $\tau = \tau_0^B$. Thus the duration of the sequential analysis cannot exceed τ_0^B.

On this basis, we can slightly weaken condition (4.33) if, as is customary in communication systems, the specified parameter is $p = \frac{1}{2}(\alpha + \beta)$, rather than α and β separately:

$$p \geq (1 + \ln S)/S. \tag{4.34}$$

The cumulative distribution function of the decision-making time τ is the probability that a nonzero number of quantum transitions will have been completed by time τ:

$$F_k(\tau) = \begin{cases} 1 - \exp(-\bar{\lambda}_k \tau), & \tau < \tau_0^B, \\ 1, & \tau \geq \tau_0^B. \end{cases} \tag{4.35}$$

Thus the probability density of τ is

$$W_k(\tau) = \frac{\partial F_k(\tau)}{\partial \tau} = \begin{cases} \bar{\lambda}_k \cdot \exp(-\bar{\lambda}_k \tau), & \tau < \tau_0^B, \\ \delta(\tau - \tau_0^B)\exp(-\tau\bar{\lambda}_k), & \tau = \tau_0^B, \\ 0, & \tau > \tau_0^B. \end{cases} \quad (4.36)$$

The mean analysis time may thus be determined directly:

$$\bar{\tau}_1^{sq} = \int_0^{\tau_0^B} W_1(\tau)\tau d\tau = \frac{1}{\bar{\lambda}_1}(1 - e^{-\bar{\lambda}_1 \tau_0^B}) \simeq \bar{\lambda}_1^{-1}, \quad (4.37)$$

and, as expected, $\bar{n}_1^{sq} = \bar{\lambda}_1 \bar{\tau}_1^{sq} \simeq 1$. Similarly,

$$\bar{\tau}_0^{sq} = \frac{-\ln \beta}{\bar{\lambda}_s} \simeq \tau_0^B, \quad (4.38)$$

corresponding to the duration of the usual, nonsequential procedure with the same β.

A brief summary of our results for systems with passive spacing now follows.

1. The sequential algorithm for signal detection (whose principles are realized in communication systems by the interrogation principle) yields a considerable reduction in energy outlay; the mean analysis time (and consequently also the mean number of interrogations) differs for different alternatives, given the same probability of error or failure to detect.

2. The lower limit of mean signal energy is one quantum, regardless of the required error probability.

3. If the background noise level is low, the sequential algorithm will be limited: its length does not exceed that of a nonsequential procedure with the same error probability. The admissible noise level depends on the magnitude of the error.

4. Wald's method of determining mean sample size yields accurate results only for high background noise levels. If the noise level is low, the technique must be refined by introduc-

ing step-function thresholds; otherwise the method leads to large errors in the figures for mean energy outlay and efficiency of sequential analysis.

B. SYSTEM WITH ACTIVE SPACING

In a system with active spacing one or more parameters of the optical signal are modulated so as to achieve orthogonal separation. For example, to transmit "0" (situation H_0) the transmitter might emit a horizontally polarized light signal of arbitrary shape, and to transmit "1" (situation H_1) — a vertically polarized signal (in the general case, of arbitrary shape). Analogous use may be made of such parameters as the pulse phase, optical carrier frequency or spatial position of the beam.

A convenient design for an optimum receiver in an LCS transmitting binary information is shown in Figure 4.7. The separator[12] D splits the received optical signal OS into two signals OS−1 and OS−2, which differ in the modulated parameter and are detected separately by photodetectors P_1 and P_2. The filter block processes the output currents λ_1 and

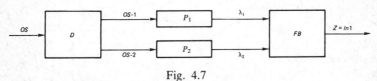

Fig. 4.7
Design for optimum receiver (active spacing):
D — separator; P_1, P_2 — photodetectors; FB — filter unit.

---

[12] The physical design of the separator depends on the modulated parameter. Thus, if the polarization is modulated, D is a polarizer; in frequency modulation, D is a system consisting of a beam splitter (semitransparent mirror) and two optical filters; if the modulated parameter is the spatial position of the beam, no separator is necessary — separation is automatic: the photodetector is placed at the points corresponding to the position of the beam under hypotheses H_0 and H_1, respectively.

λ_2 of the detectors and computes the logarithmic likelihood ratio $z = \ln \Lambda$. The decision as to which signal was transmitted is made by comparing z with the threshold.

B.1. *Formulation of the problem.* Since the Poisson distribution, which governs the photodetector currents, is uniquely determined by its expectation, we may formulate the problem of identifying two signals (generally nonsteady) in terms of statistical decision theory in the following way.

Hypothesis H_1 is that the first moment functions of the photocurrents of detectors P_k ($k = 1, 2$) are

$$\bar{\lambda}_k(t) = \bar{\lambda}_{k1}(t) \qquad (k = 1, 2),$$

against the single alternative H_0:

$$\bar{\lambda}_k(t) = \bar{\lambda}_{k0}(t) \qquad (k = 1, 2).$$

This involves simultaneous testing of hypotheses concerning the parameters of *two* Poisson distributions (if the number of transmitted signals is N, the problem will be simultaneous testing of hypotheses about the parameters of N distributions).

On this basis, the logarithm of the likelihood ratio may be expressed as the sum of two expressions of type (4.15):

$$z = \ln \Lambda = \int_0^\tau \lambda_1(t) \ln \frac{\bar{\lambda}_{11}(t)}{\bar{\lambda}_{10}(t)} dt - \int_0^\tau [\bar{\lambda}_{11}(t) - \bar{\lambda}_{10}(t)] dt$$

$$+ \int_0^\tau \lambda_2(t) \ln \frac{\bar{\lambda}_{21}(t)}{\bar{\lambda}_{20}(t)} dt - \int_0^\tau [\bar{\lambda}_{21}(t) - \bar{\lambda}_{20}(t)] dt$$

$$+ \ln \frac{P(H_1)}{P(H_0)}, \qquad (4.39)$$

where $\lambda_k(t)$, $k = 1, 2$, are samples from the output of detector P_k (in photoelectrons per second), $\bar{\lambda}_{ki}(t)$ is the first moment function of $\bar{\lambda}_k$ in the event that H_i is true, and $P(H_i)$, $i = 0, 1$, is the prior probability of H_i.

The combination of Eqs. (4.39), (4.16) and (4.17) completely define a sequential algorithm.

Sec. 4.2. MAXIMUM NOISE IMMUNITY OF FEEDBACK SYSTEMS

We shall assume henceforth that the signals are equiprobable, symmetric and steady:

$$P(H_0) = P(H_1),$$
$$\bar{\lambda}_{10}(t) = \bar{\lambda}_{21}(t) \equiv \bar{\lambda}_0, \quad (4.40)$$
$$\bar{\lambda}_{11}(t) = \bar{\lambda}_{20}(t) \equiv \bar{\lambda}_1,$$

where $\bar{\lambda}_0$ is the mean background noise photocurrent, $\bar{\lambda}_s$ the mean signal photocurrent,

$$\bar{\lambda}_1 = \bar{\lambda}_0 + \bar{\lambda}_s.$$

Using (4.40), we simplify (4.39) as follows:

$$z = \ln(S+1) \int_0^\tau [\lambda_1(t) - \lambda_2(t)] dt, \quad (4.41)$$

where $S = \bar{\lambda}_s/\bar{\lambda}_0$ is the signal/noise ratio.

We define the auxiliary parameter

$$n = \int_0^\tau [\lambda_1(t) - \lambda_2(t)] dt \quad (4.42)$$

— the cumulative difference of the detector photocurrents. Then the sequential algorithm for separation of two signals may be written, via (4.16), (4.17) and (4.42), as

$$G_1: n \geq \frac{\ln A}{\ln(S+1)} = \frac{\ln(1-\beta) - \ln \alpha}{\ln(S+1)} = A' \to H_1;$$

$$G_2: n \leq \frac{\ln B}{\ln(S+1)} = \frac{\ln \beta - \ln(1-\alpha)}{\ln(S+1)} = B' \to H_0; \quad (4.43)$$

$$C: B' < n < A' \to \tau \uparrow.$$

Note that, in contrast to the treatment of passive spacing, the thresholds used here are constants, and the analysis procedure does not terminate for any S.

We shall determine the characteristics of the untruncated sequential algorithm (4.43): mean analysis time, mean signal energy, and distribution function of analysis time.

B.2. *Determination of characteristics of sequential algorithm.* We first find the mean analysis time, using Eqs. (4.19). In our case, it follows readily from (4.41) and (4.40) that

$$\mathbf{E}_k(z_i) = \bar{\lambda}_s \ln(S+1) \cdot \Delta t. \qquad (4.44)$$

Substituting (4.44) into (4.19) and multiplying by Δt, we obtain the expected analysis time:

$$\bar{\tau}_0^{sq} = \frac{\alpha \ln A + (1-\alpha)\ln B}{\bar{\lambda}_s \ln(S+1)},$$

$$\bar{\tau}_1^{sq} = \frac{(1-\beta)\ln A + \beta \ln B}{\bar{\lambda}_s \ln(S+1)}. \qquad (4.45)$$

Two conclusions follow easily from (4.54):

1) If $\alpha, \beta \ll 1$, the mean analysis time is determined by the point of intersection of the straight line representing the expectation of the sample, $\bar{n}_k(\tau) = (-1)^{k+1}\bar{\lambda}_s\tau$ $(k = 0, 1)$, with the threshold lines A' and B'.

2) If $\alpha = \beta = p \ll 1$ the mean analysis time is the same for both hypotheses:

$$\bar{\tau}^{sq} = \frac{-\ln p}{\bar{\lambda}_s \ln(S+1)}. \qquad (4.46)$$

This equation yields the mean energy requirements in both situations:

$$\bar{n}^{sq} = \bar{\lambda}_s \cdot \bar{\tau}^{sq} = \frac{-\ln p}{\ln(S+1)}. \qquad (4.47)$$

It remains to find the distribution function of the analysis time. Neglecting double excess over the thresholds (i.e., ignoring the case in which n attains first the upper and then the lower limit, or *vice versa*), we determine the cumulative distribution function as follows [26]:

$$F(\tau) = 1 - \sum_{n=-n_0}^{n_0} e^{-[1+(S+1)^{-1}]\bar{\lambda}_1\tau} \cdot (S+1)^{n/2} \cdot I_n\left(\frac{2\bar{\lambda}_1\tau}{\sqrt{S+1}}\right),$$

$$(4.48)$$

Sec. 4.2. MAXIMUM NOISE IMMUNITY OF FEEDBACK SYSTEMS

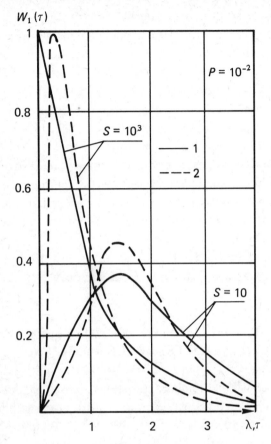

Fig. 4.8
Distribution function of analysis time for active spacing ($p = 10^{-2}$; $S = 10$ and $S = 10^3$):
1) computed from exact formulas 2) Wald distribution.

where $n_0 = [A']^{13}$ and $J_n(\ )$ is the n-th order Bessel function of an imaginary argument.

The probability density, readily found by differentiating (4.48) with respect to τ, is plotted in Figure 4.8. For comparison, the figure also shows the corresponding Wald distribution functions (dashed curves).

---

[13] $[x]$ denotes the integral part of the number x.

We have examined sequential algorithms for reception with both passive and active spacing, determining the noise immunity of feedback systems with memory in optimum operating conditions — the maximum noise immunity feasible in LCS with feedback.

It is now possible to compare the merits of active versus passive spacing. Comparison of Eqs. (4.46), (4.47) with the corresponding equations for passive spacing shows that active spacing yields a savings in mean analysis time, as against a loss in mean energy outlay (taking the mean outlay for each hypothesis into account). In accordance with our criteria, then, we shall henceforth confine attention to systems with passive spacing.

We now proceed to the simpler (and therefore less immune to noise) systems without memory.

4.3. NOISE IMMUNITY OF MEMORYLESS SYSTEMS WITH FEEDBACK.[14] FEEDBACK EFFICIENCY

We now consider the characteristics of a memoryless quantum system which transmits binary information with symbol-by-symbol interrogation [18]. We shall assume that the indeterminate region B and the acceptance regions G_k ($k = 0, 1$) do not change from one repetition to the next (the effect of bounds imposed on the number of repetitions will be considered in §4.5). As before, the received signal will be a Poisson stream of quantum transitions.

We shall assume that the feedback path is error-free (the effect of the feedback path on system performance will be examined in §4.5) and, in accordance with the results of §4.2, the analysis will be concerned only with systems with passive spacing.

---

[14] In memoryless systems, the decisions of the forward-path receiver are based solely on the information conveyed by the last elementary pulse relating to the message element in question.

In addition, the following assumptions will be adopted: 1) Both message elements are equiprobable and of equal cost. 2) The performance indexes are the error probability and the mean energy requirements. 3) The transmitter emits pulses of fixed duration and energy.

Under these assumptions, we now proceed to determine the noise immunity of memoryless systems and optimum algorithms for their operation.

As in the preceding section, the receiver in the forward path divides the signal space into acceptance regions G_k and an indeterminate region (or null zone) B. If the signal lies in region G_k, hypothesis H_k is accepted, and the transmitter is notified that a decision has been made and transmission of the element may be stopped. But if the signal lies in the null zone, no final decision is made and the transmitter is requested to send additional information about the same element.

We define the regions G_k and B as follows. Suppose that the hypothesis H_0 corresponds to a passive pulse (only noise energy in the channel), H_1 to an active pulse (signal and noise in the channel), and that at each step the receiver computes the logarithm of the likelihood ratio:

$$\ln \Lambda = n \cdot \ln \left(1 + \frac{\bar{\lambda}_s}{\bar{\lambda}_0}\right) - \bar{n}_s, \qquad (4.49)$$

where n is the number of quantum transitions during the observation period, $\bar{n}_s$ the mean number of transitions due to a signal during the observation period (signal energy), $\bar{\lambda}_s$ and $\bar{\lambda}_0$ the mean frequencies of transitions (mean number of transitions per unit time) due to signal and background noise, respectively.

In this context, the "observation period" is the duration of the one incoming elementary pulse in a memoryless system, and the duration of this pulse and all the preceding pulses relating to the same message element in a system with memory.

Suppose that the receiver compares Λ with symmetric threshold values of the likelihood ratio (Λ_0 and $1/\Lambda_0$). Then, in view of (4.49), the regions G_k and B are defined by the following formulas:

$$G_1: n \geq \frac{\bar{n}_s + \ln \Lambda_0}{\ln(S+1)} = a,$$

$$G_0: n \leq \frac{\bar{n}_s - \ln \Lambda_0}{\ln(S+1)} = b, \qquad (4.50)$$

$$B: b < n < a,$$

where Λ_0 is the threshold value of the likelihood ratio and $S = \bar{\lambda}_s/\bar{\lambda}_0$ the signal/noise ratio in the channel.

Using (4.50) and the general formulas of [18], we obtain the following relations for memoryless systems:

$$\alpha = \frac{1 - F(a, \bar{n}_{ep}/S)}{1 + F(b, \bar{n}_{ep}/S) - F(a, \bar{n}_{ep}/S)},$$

$$\beta = \frac{F(b, \bar{n}_{ep} + \bar{n}_{ep}/S)}{1 + F(b, \bar{n}_{ep} + \bar{n}_{ep}/S) - F(a, \bar{n}_{ep} + \bar{n}_{ep}/S)}, \qquad (4.51)$$

$$\bar{n} = \frac{\bar{n}_{ep}}{1 + F(b, \bar{n}_{ep} + \bar{n}_{ep}/S) - F(a, \bar{n}_{ep} + \bar{n}_{ep}/S)}$$

where $\alpha = \mathbf{P}(H_1|H_0)$ and $\beta = \mathbf{P}(H_0|H_1)$ are the error probabilities, $F(x, y) = \sum_{i=0}^{x} (y^i/i!)e^{-y}$ is the Poisson cumulative distribution function, $\bar{n}_{ep}$ is the signal energy per elementary pulse (in the forward path), and $\bar{n}$ is the mean signal energy per unit transmitted information (also in the forward path). Note that the factor $\frac{1}{2}$, corresponding to the equal prior probabilities of the alternatives, has been omitted from the formula for $\bar{n}$ (and will be omitted throughout the sequel; the reason for the appearance of this factor is that no energy is used when the alternative is H_0).

Since the received samples are discrete, the numbers a and b given by (4.50) should be rounded off to the nearest integer from below before substitution into (4.51); as a result, the algorithm is no longer symmetric with respect to the error probabilities: $\alpha \neq \beta$. The computation shows, however, that the asymmetry is relatively insignificant: if $S \leq 100$, then $1/3 \leq \alpha/\beta \leq 3$.

Since, as follows from (4.50), the thresholds a and b depend on the channel characteristics and on Λ_0, Eqs. (4.51) may be treated as parametric equations for the noise immunity curves:

$$p = f_1(S, \bar{n}_{ep}, \Lambda_0),$$
$$\bar{n} = f_2(S, \bar{n}_{ep}, \Lambda_0),$$
(4.52)

where $p = (\alpha + \beta)/2$ is the mean error probability and Λ_0 acts as a parameter.

Fixing S, assigning different values to $\bar{n}_{ep}$ and varying Λ_0, we obtain the mean error probability p as a function of the mean signal energy, shown by solid curves in Figures 4.9a ($S = 1$), 4.9b ($S = 4$) and 4.9c ($S = 10, \infty$).

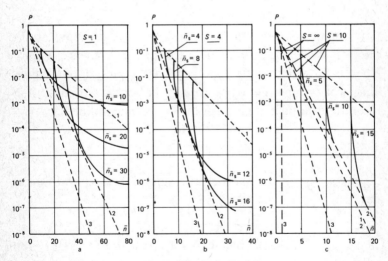

Fig. 4.9

Noise immunity of different optical communication systems (passive spacing, nonfluctuating signal)

solid curves — nonoptimized system with feedback, without memory; dashed curves — optimized systems: 1) without feedback; 2) with feedback, without memory; 3) with feedback and memory (sequential analysis).

Ch. 4. NOISE IMMUNITY OF LCS

Setting $\Lambda_0 = 1$ ($a = b$) in (4.51) (signifying reception without an indeterminate region), we obtain the equations of the noise immunity curves for one-way systems, $p_0 = f(\bar{n}, S)$; clearly, each of the noise immunity curves (4.52) "begins" from one of the curves $p_0 = f(n, S)$ (as we can always put $\Lambda_0 = 1$ in (4.52)). This is clearly evident in Figure 4.9, where the curves $p_0 = f(n, S)$ are dashed (curves 1). It is also clear from Figure 4.9 that, just as in the case of normally distributed fluctuation noise [18], at small S memoryless systems are more efficient than one-way systems only for bounded values of p.

$$p_0 > p > p_1. \tag{4.53}$$

For example, if $S = 1$,
for $\bar{n}_{ep} = 10 : p_0 = 0.9 \cdot 10^{-1}, p_1 = 10^{-3}$;
for $\bar{n}_{ep} = 20 : p_0 = 3.0 \cdot 10^{-2}, p_1 = 10^{-5}$;
for $\bar{n}_{ep} = 30 : p_0 = 1.1 \cdot 10^{-2}, p_1 = 10^{-7}$.

At $S > 3$, however, memoryless systems show superior performance in any mode of operation as compared with one-way systems. Moreover, at $S > 10$ it is always advantageous to enlarge the indeterminate region;[15] put differently, for optimum performance in the sense of minimum $\bar{n}$, the indeterminate region should be of maximum size:

$$B_{max}: b = 0, \quad a = \frac{2\bar{n}_{ep}}{\ln(S+1)}. \tag{4.54}$$

Hence a simple parametric representation of the optimum performance curves when $S > 10$:

$$p = \frac{\exp(-\bar{n}_{ep} - \bar{n}_{ep}/S)}{1 - F(2\bar{n}_{ep}/\ln(S+1); \bar{n}_{ep} + \bar{n}_{ep}/S)},$$

$$\bar{n} = \frac{\bar{n}_{ep}}{1 - F(2\bar{n}_{ep}/\ln(S+1); \bar{n}_{ep} + \bar{n}_{ep}/S)}. \tag{4.55}$$

---

[15] Strictly speaking, the ranges of S values in which these statements are true depend (albeit weakly) on the considered interval of error probabilities. The figures 3 and 10 in the text correspond to a wide range of error probabilities — from 0.5 to 10^{-8}.

Parametric equations of these curves for $S < 10$ may be obtained using Lagrange multipliers [18]; this method is of course more general (but also more complicated).

An examination of the optimum curves in Figure 4.9 (dashed curves, curves 2) leads to the following conclusions. The noise immunity of a memoryless system in optimum operating modes is practically independent of the signal/noise ratio, over a fairly wide range of values; as S falls from ∞ to 10, the mean energy required to achieve error probability 10^{-6} rises by only 0.41 db. For comparison, we note that the appropriate figure for a one-way system is 4.11 db, and for optimum systems with memory 9.03 db (see §4.2). The optimum noise immunity curves for systems with memory (noise immunity of a sequential detection algorithm; see §4.2) are also shown in Figure 4.9 (dashed curves 3).

In §4.2 we defined the energy efficiency V as the ratio of the mean energy required to achieve prescribed error probability in a one-way system to the corresponding quantity in a system with interrogation. Corresponding curves $V(S, p)$ are plotted in Figure 4.10 for systems with and without memory. One sees from the figure that systems with memory yield the best results in the absence of noise; increased noise tends to equalize the performance of the two types of system, as is generally the case when the fluctuation noise level rises.

The behavior of the function $V(S)$ is slightly more complicated in memoryless systems. As $S \to \infty$, the optimum threshold of a one-way system approaches zero; thus the indeterminate region becomes smaller and as a result the characteristics of memoryless systems resemble those of one-way systems at large S values. The fall of V when S drops to less than $S = 10$ is due to the above-mentioned equalization of system performance at high fluctuation noise levels. We recall that this is just the threshold value of S at which an indeterminate region of maximum size ceases to be optimal.

Ch. 4. NOISE IMMUNITY OF LCS

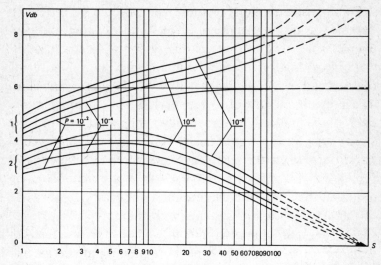

Fig. 4.10
Dependence of feedback efficiency on background noise level: 1) optimized systems with memory (sequential analysis), 2) optimized systems without memory.

A similar equalization holds for systems with and without memory at high noise levels: at $S = 1$ their noise immunity differs by at most 1.7 db.

We can summarize the above conclusions from the data of Figure 4.10 (the curves of $V(S)$) as follows:

a) At large signal/noise ratios ($S > 30$), feedback systems with memory provide superior immunity to noise, though technically more complex.

b) At medium S values (from 1 to 30) memoryless systems are preferable, since they are only slightly inferior in performance to systems with memory, while always much simpler in technical implementation.

c) At small S ($S < 1$), neither type of system is very efficient.

The analysis of §§4.2 and 4.3 has thus produced clear-cut conclusions as to the most advantageous type of feedback system, depending on channel noise. One factor we have ignored, however, is the effect of signal fluctuations. In systems with open-air beam propagation these fluctuations are due mainly to turbulent irregularities in the atmosphere, and they may be considerably deleterious for the noise immunity of an LCS. This question will be considered in the next section.

4.4. OPTICAL SIGNAL FLUCTUATIONS IN A TURBULENT ATMOSPHERE. FEEDBACK EFFICIENCY

The intensity of an optical signal travelling through a turbulent atmosphere is a random variable whose distribution depends on the length of its track and the aperture of the receiving antenna. We have already stated (§2.2) that in most cases (if the track is not too long and the aperture relatively narrow) the intensity obeys a log-normal distribution law [25].

The frequency spectrum of the signal fluctuations presents a rather more difficult problem. We shall assume that the signal intensity varies so slowly that it may be considered constant 1) during the transmission of each elementary pulse and even 2) during the transmission of each message element (including all possible repetitions).[16]

Under this assumption, $\bar{n}_{ep}$ — the mean number of signal photoelectrons ejected from the photodetector during one elementary pulse — is a random variable governed (thanks to the first part of the assumption) by the same distribution as the signal intensity:

$$W(\bar{n}_{ep}) = \frac{1}{\sigma\sqrt{2\pi}\bar{n}_{ep}} \exp\left[-\frac{\ln^2\left(\frac{\bar{n}_{ep}}{\bar{\bar{n}}_{ep}}\exp(\sigma^2/2)\right)}{2\sigma^2}\right],$$
(4.56)

---

[16] This assumption is generally valid for laser communication lines with high data rates.

where σ^2 is the variance of the log-intensity,

$\bar{n}_{ep}$ is the mean value of $\bar{n}_{ep}$.

By virtue of the second part of the assumption, we may assume that all equations obtained in §§4.2 and 4.3 for constant-parameter systems[17] (the formulas for $\bar{n}$, p, etc.) have the same form here, except that the parameters they define are now random variables. System performance will now depend on the expectations of these parameters.

Thus, if the characteristic error probabilities of a constant-parameter system are α and β, and the mean energy outlay per unit transmitted data is $\bar{n}$, then the corresponding characteristics of an analogous system with random $\bar{n}_{ep}$ are defined by

$$\tilde{\alpha} = \int_0^\infty \alpha \cdot W(\bar{n}_{ep}) d\bar{n}_{ep},$$

$$\tilde{\beta} = \int_0^\infty \beta \cdot W(\bar{n}_{ep}) d\bar{n}_{ep}, \qquad (4.57)$$

$$\tilde{n} = \int_0^\infty \bar{n} \cdot W(\bar{n}_{ep}) d\bar{n}_{ep},$$

where $\tilde{\alpha}$, $\tilde{\beta}$, $\tilde{n}$ are the expectations of α, β, $\bar{n}$ and $W(\bar{n}_{ep})$ is defined by (4.56).

We now turn to memoryless systems with interrogation, retaining all the assumptions adopted in §4.3 for such systems. The parameters α, β, $\bar{n}$ are defined by (4.51) and, according to (4.57), the characteristics of the system in the new situation may be found by averaging (4.51) with the weight function (4.56). In computing the appropriate integrals, the following points should be remembered:

1) In the computation of $\tilde{n}$, the value of $\bar{n}_{ep}$ appearing in the formula for $\bar{n}$ should be considered a constant, equal to $\bar{n}_{ep}$, since the transmitter output consists of pulses of fixed energy.

---

[17] I.e., systems with nonfluctuating signals.

2) The quotient $\tilde{n}_{ep}/S$ appearing in all of Eqs. (4.51) and defining the intensity of background fog should also be treated as a constant $\tilde{n}_{ep}/S$, since the noise level does not usually vary in the course of one communication session (though it does display very slow deviations from its mean value).

3) The thresholds a and b appearing in (4.51) may either vary with $\bar{n}_{ep}$ (adaptive system) or remain constant, but even in the latter case they may not coincide with the symmetric thresholds defined by (4.50).

With all these remarks in mind, substitution of (4.51) and (4.56) into (4.57) for a nonadaptive optical communication system gives

$$\tilde{\alpha} = \frac{1 - F(a, \tilde{n}_{ep}/S)}{1 + F(b, \tilde{n}_{ep}/S) - F(a, \tilde{n}_{ep}/S)},$$

$$\tilde{\beta} = \frac{1}{\sigma\sqrt{2\pi}} \times$$

$$\times \int_0^\infty \frac{F(b, \bar{n}_{ep} + \tilde{n}_{ep}/S) \exp\left[-\dfrac{\ln^2\left(\dfrac{\bar{n}_{ep}}{\tilde{n}_{ep}} \exp \dfrac{\sigma^2}{2}\right)}{2\sigma^2}\right]}{\bar{n}_{ep}[1 + F(b, \bar{n}_{ep} + \tilde{n}_{ep}/S) - F(a, \bar{n}_{ep} + \tilde{n}_{ep}/S)]} d\bar{n}_{ep},$$

$$\tilde{n} = \frac{\tilde{n}_{ep}}{\sigma\sqrt{2\pi}} \int_0^\infty \frac{\exp\left[-\dfrac{\ln^2\left(\dfrac{\bar{n}_{ep}}{\tilde{n}_{ep}} \exp \dfrac{\sigma^2}{2}\right)}{2\sigma^2}\right]}{\bar{n}_{ep}[1 + F(b, \bar{n}_{ep} + \tilde{n}_{ep}/S) - F(a, \bar{n}_{ep} + \tilde{n}_{ep}/S)]} d\bar{n}_{ep}$$

(4.58)

(if the prior probabilities of the hypotheses are equal, $p = \frac{1}{2}(\tilde{\alpha} + \tilde{\beta})$).

The noise immunity curves of Figure 4.11 were derived from Eqs. (4.58) for an optimized system, using a computer; in view of the third of the above remarks, the optimum thresholds a and b were determined numerically. It turns out that, as in a constant-parameter system, the optimum thresholds are symmetric with respect to d — the optimum

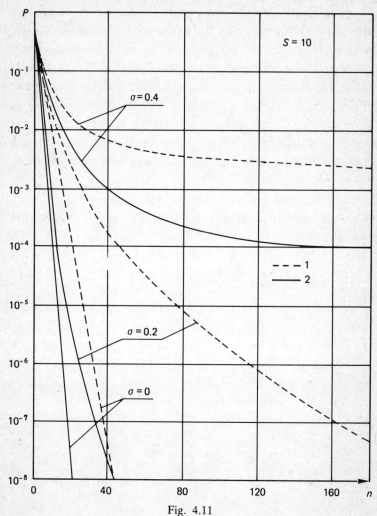

Fig. 4.11

Noise immunity of LCS with feedback in presence of atmospheric fluctuations of optical signal (system without memory, number of repetitions unlimited):

1) system without feedback; 2) optimized system with feedback, without memory.

threshold in a one-way system. However, d itself is 10% less if $\sigma = 0.2$ and 20% less if $\sigma = 0.4$.

For comparison of systems with interrogation to one-way systems as regards energy requirements, we again use the energy efficiency V (see above). The $V(\sigma)$ curves plotted in Figure 4.12 show that systems with interrogation in a random-parameter channel are more efficient (sometimes considerably more efficient) than similar systems in constant-parameter channels. Moreover, it is evident from Figure 4.11 that in a highly turbulent atmosphere the use of interrogation enables the error probability to be lowered to a level practically unattainable in one-way systems.

We have thus ascertained the effect of a turbulent atmosphere on the noise immunity and efficiency of LCS with feedback. The data justify the assertion that it is precisely under turbulent atmospheric conditions that feedback is most beneficial.

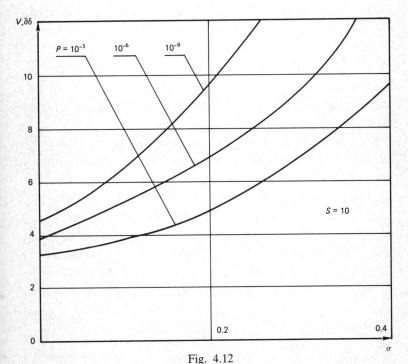

Fig. 4.12

Dependence of feedback efficiency on fluctuations of optical signal (computed for optimized system without memory, unlimited number of repetitions).

4.5. EFFECT OF FEEDBACK PATH AND LIMITED NUMBER OF REPETITIONS ON FEEDBACK EFFICIENCY

So far, we have completely ignored questions connected with the feedback path (assumed error-free) and the number of repetitions (assumed unbounded). In real systems, however, errors in the feedback path may produce new errors in the forward path; and the bounded number of repetitions, unavoidable for technical reasons, invariably impair the efficiency of feedback systems.[18] In addition, the cost of installing the feedback path[19] may make the very idea of introducing feedback less attractive than supposed.

These considerations are physically self-evident; we shall now try to interpret them from a quantitative point of view.

As before, we consider an optical communication system transmitting binary data with symbol-by-symbol interrogation, assuming that the indeterminate region B and acceptance regions G_k ($k = 0, 1$) remain unchanged throughout all repetitions.

We adopt the following assumptions: 1) the two message elements transmitted along the forward path are equiprobable and of equal cost; 2) each signal is allotted a specific time interval for transmission in both paths (for example, the receiver at the end of the feedback path "knows" the position in time of every possible interrogation signal relating to any message element); 3) both transmitters operate in the passive-spacing mode, producing monochromatic, coherent light pulses of fixed duration and energy; 4) the additive

---
[18] In comparison with the limiting values found in §§4.2, 4.3 and 4.4.
[19] The word "cost" is used here in a very wide sense, including the actual cost of setting up the path (if this is done specially), the cost of extra equipment at both ends of the line, the energy required by the transmitter in the feedback path, etc.

Sec. 4.5. FEEDBACK PATH AND LIMITED REPETITIONS 107

noise level in both paths is low in comparison with the signal level; 5) both receivers employ direct detectors.

The notation will be as follows:

n — number of photoelectrons ejected from detector per elementary pulse;

n_{ij} — mean value of n for i-th path when hypothesis H_j is true (i.e., energy registered by the photodetector per elementary pulse); $i = 0$ indicates the forward path, $i = 1$ the feedback path; when $i = 1$, $j = 0$ signifies the "decision taken" signal, $j = 1$ the "interrogation" signal.

$\bar{n}_{ep} = \bar{n}_{01} - \bar{n}_{00}$ — signal energy per elementary pulse in forward path;

$S = \bar{n}_{ep}/\bar{n}_{00}$ — signal/noise ratio in forward path;

$\mathbf{P}(H_k|H_p)$ — probability of accepting decision H_k with hypothesis H_p true in the forward path;

$= (\mathbf{P}(H_0|H_1) + \mathbf{P}(H_1|H_0))/2$ — mean error probability in forward path;

$\mathbf{P}_{ij}(n=0|H_p)$ — probability that upon transmission of the j-th elementary pulse no photoelectron is ejected ($n = 0$) from the detector of the i-th path, given that H_p is true;

$\mathbf{P}_i(n>0|H_p)$ — probability of opposite event;

$\mathbf{P}_i(j|j-1)$ — probability that transmitter of i-th path will be requested to transmit a j-th elementary pulse, given that the $(j-1)$-th pulse was requested; $j \leq k$, where k is the maximum number of repetitions (counting the first elementary pulse);

$\bar{k}_j$ — mean number of repetitions when H_j is true;

$\bar{\bar{n}}_{ij} = \bar{k}_j \bar{n}_{ij}$ — mean energy registered in i-th path when H_j is true.

Note that by assumption 4 the additive noise level in both paths is negligibly low compared with the signal level[20] and $\bar{n}_{ii} = 0$.

---

[20] It is interesting that in this case the information feedback system of [14] is not superior in performance to a one-way system. This is due to the formulation of the problem in [14]: the authors essentially consider a problem of parameter estimation rather than data transmission.

Under these conditions, the time required for sequential analysis of signals with passive spacing does not exceed that required for the usual (nonsequential) procedure with the same error probability; and the signal is assigned to the indeterminate region if and only if the number of photoelectrons is zero (§4.2). In other words, in the forward receiver of a system with memory and interrogation, decision H_1 is taken after registration of the first photoelectron; if transmission of the q-th elementary pulse produces no photoelectrons, an interrogation signal is sent if $q < k$, and decision H_0 is taken if $q = k$. This decision-making scheme is also suitable for a memoryless system in which the acceptance regions vary from repetition to repetition.

Note that in view of the low noise level and the assumption that the detector current is Poisson-distributed, we have

$$\mathbf{P}_{ij}(n = 0 | H_i) = \exp(-\bar{n}_{il}), \quad l \neq i$$
$$\mathbf{P}_{ij}(n > 0 | H_i) = 0, \quad i = 0, 1, \quad j \leq k. \tag{4.59}$$

Therefore, since the feedback receiver is assumed to be operating with zero threshold, the only effect of the fact that the feedback path is nonideal is to enlarge the mean number of repetitions (and hence also the mean energy requirements in the forward path); the error probabilities will be the same as in error-free operation of the feedback path:

$$\mathbf{F}(H_1 | H_0) = 0;$$

$$p = \frac{1}{2} \mathbf{P}(H_0 | H_1) = \frac{1}{2} \prod_{j=1}^{k} \mathbf{P}_{0j}(n = 0 | H_1) = \frac{1}{2} \prod_{j=1}^{k} e^{-\bar{n}_{01}} = \frac{1}{2} e^{-k \cdot \bar{n}_{01}}. \tag{4.60}$$

Hence it follows that

$$\bar{n}_{01} = -\frac{\ln 2p}{k}. \tag{4.61}$$

In order to lower the energy requirements in the forward path, the feedback loop may be checked by stipulating that, even after a decision H_1 has been made, the forward receiver continues to analyze all the remaining elementary pulses, up to and including the k-th, and whenever $n \neq 0$ repeats the "decison taken" signal. Under this assumption, the mean energy requirements in the forward path for message element will be

$$\bar{\bar{n}}_{01} = \sum_{j=i}^{k} \bar{n}_{01} \prod_{i=1}^{i} \mathbf{P}_0(i|i-1) = \bar{n}_{01} \cdot \frac{1 - \mathbf{P}_0^k(j|j-1)}{1 - \mathbf{P}_0(j|j-1)}, \quad (4.62)$$

where

$$\mathbf{P}_0(j|j-1) = \mathbf{P}_{oj}(n=0|H_1) + \mathbf{P}_{oj}(n>0|H_1) \cdot \mathbf{P}_{1j}(n=0|H_0) \quad (4.63)$$

$$= \exp(-\bar{n}_{01}) + [1 - \exp(-\bar{n}_{01})] \exp(-\bar{n}_{10}).$$

Recall that $\bar{n}_{ii} = 0$, so that $\bar{\bar{n}}_{ii} = 0$; $\bar{n}_{01}$ may be derived from the prescribed error probability using Eq. (4.61).

To determine $\bar{n}_{10}$, let us assume that every error in the feedback path is detected immediately (i.e., during transmission of the next pulse along the forward path): $\mathbf{P}_{oj}(n=0|H_1) \ll 1$. Then

$$\bar{\bar{n}}_{10} = \sum_{i=0}^{k-1} \bar{n}_{10} \prod_{j=1}^{i} \mathbf{P}_{1j}(n=0|H_0) = \bar{n}_{10} \cdot \frac{1 - \mathbf{P}_{1j}^k(n=0|H_0)}{1 - \mathbf{P}_{1j}(n=0|H_0)} \quad (4.64)$$

$$= \bar{n}_{10} \cdot \frac{1 - e^{-k\bar{n}_{10}}}{1 - e^{-\bar{n}_{10}}}.$$

To assess the energy requirements for the entire system, we must define some kind of equivalent energy $\bar{\bar{n}}$. The most natural definition makes $\bar{n}$ a linear function of the $\bar{\bar{n}}_{ij}$'s, defined in terms of a payoff matrix c_{ij} (where c_{ij} represents the cost of $\bar{\bar{n}}_{ij}$):

$$\bar{n} = \sum_i \sum_j C_{ij} \bar{\bar{n}}_{ij} = C_{01} \cdot n_{01} + C_{10} \cdot \bar{\bar{n}}_{10}.$$

Without loss of generality, we may assume that one of the

110 Ch. 4. NOISE IMMUNITY OF LCS

c_{ij}'s, say c_{01}, is unity, so that

$$\bar{\bar{n}} = \bar{\bar{n}}_{01} + c_{10} \cdot \bar{\bar{n}}_{10} = \bar{\bar{n}}_{01} + c \cdot \bar{\bar{n}}_{10}, \qquad (4.65)$$

where $\bar{\bar{n}}_{01}$ and $\bar{\bar{n}}_{10}$ are defined by (4.62) and (4.64), respectively. Hence forth we shall drop the subscript of c_{10}, denoting it simply by c.

We now have all the tools to determine the mean energy in the forward path (4.62) and feedback path (4.64), and the equivalent energy requirements (4.65) for the system as a whole. Each of these quantities depends on the maximum number k of repetitions (Figure 4.13). We shall confine attention to the equivalent energy $\bar{\bar{n}}$, as it includes both $\bar{\bar{n}}_{01}$ and $\bar{\bar{n}}_{10}$ (to obtain these we need only set $c = 0$ and $c = 1$, respectively, in the equation for $\bar{\bar{n}}$). It is readily seen that n is a monotonically decreasing function of k, and

$$\bar{\bar{n}}_{min} = \lim_{k \to \infty} \bar{\bar{n}} = \frac{1 - p^{1-\exp(-\bar{n}_{10})} + c\bar{n}_{10}}{1 - \exp(-\bar{n}_{10})}. \qquad (4.66)$$

Figure 4.14 shows a plot of $\bar{\bar{n}}_{min}$ as a function of the energy $\bar{\bar{n}}_{10}$

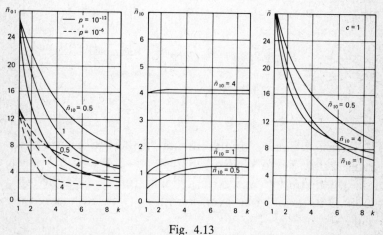

Fig. 4.13
Dependence of average energy requirements in system with feedback on maximum number of repetitions (background level assumed low):
a) energy requirements in forward path; b) energy requirements in feedback path; c) total energy requirements.

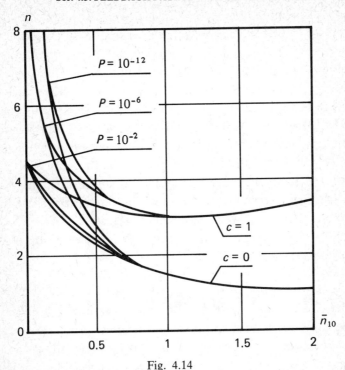

Fig. 4.14
Dependence of average energy requirements in system with feedback on energy per elementary pulse in feedback path (background level assumed low, number of repetitions unlimited):
1) energy requirements in forward path, 2) total energy requirements.

per elementary pulse in the feedback path. It may be seen that $\bar{\bar{n}}_{min}$ is not a monotonic function of $\bar{n}_{10}$;[21] it reaches a minimum at a point $\bar{n}_{10}^{opt}$ defined by

$$\left.\frac{\partial \bar{\bar{n}}_{min}}{\partial \bar{n}_{10}}\right|_{\bar{n}_{10}=\bar{n}_{10}^{opt}} = 0. \qquad (4.67)$$

For small p, substitution of (4.66) into (4.67) yields a transcendental equation for $\bar{n}_{10}^{opt}$:

$$C \cdot \exp \bar{n}_{10}^{opt} = 1 + c + c \cdot \bar{n}_{10}^{opt}. \qquad (4.68)$$

---

[21] If $c = 0$, n_{10} may always be increased. However, n_{min} approaches its limiting value (1 quantum) quite rapidly, so that one readily determines the value n_{10} corresponding to a practically error-free feedback path.

112   Ch. 4. NOISE IMMUNITY OF LCS

Substituting the solutions of this equation into (4.66), we obtain $\bar{\bar{n}}_{\min}^{\min}$ — the minimum possible energy requirements in the system. Surprisingly, $\bar{\bar{n}}_{\min}^{\min}$ is finite even when $p \to 0$. Thus, if $c = 0$ we have $\bar{\bar{n}}_{\min}^{\min} \to 1$ as $p \to 0$ (in fact, we proved this in §4.2). But if $c = 1$, then $\bar{\bar{n}}_{\min}^{\min} = 3.15$. These two special cases correspond to two situations of practical importance: 1) energy consumption in the feedback path is insignificant; and 2) energy in the feedback path costs the same as in the forward path. The first situation obtains, for example, in transmission from on board a moving object, the second in data exchange between equivalent objects.

Finally, we consider the energy efficiency of an optimum feedback system:

$$V = \bar{\bar{n}}/\bar{\bar{n}}_{\min}^{\min}, \qquad (4.69)$$

where $\bar{\bar{n}} = -\ln 2p$ is the mean energy required to achieve the same error probability p in a one-way system.

Curves of V versus c for different p values are shown in Figure 4.15. An interesting observation from the figure is

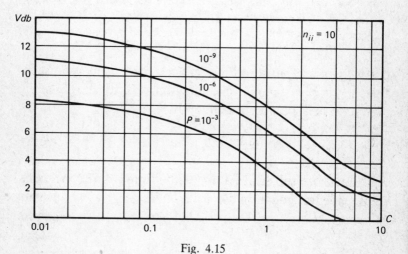

Fig. 4.15
Dependence of feedback efficiency on cost of energy in feedback path (background noise level assumed low, unlimited number of repetitions, system optimized).

that even when $c>1$ in an optimum system, we have $V_{db}>0$ (and it may be quite high). In other words, it may be advantageous to use feedback systems even when energy costs more in the feedback path than in the forward path. This fact notwithstanding, the efficiency of the system is of course higher, the lower the cost of energy in the feedback path.

CONCLUSIONS

We now summarize the main findings of our study of the various ways of introducing feedback in laser communication systems (in particular, the possibility of interrogation) and of optimizing such systems.

1. Time division multiplexing is preferable to frequency division multiplexing.

2. The correlation between signal/noise ratio and the preferable type of system is as follows:

large S ($S>30$): feedback systems with memory;
medium S ($1<S<30$): memoryless systems;
small S ($S<1$): neither type of system is efficient.

3. At sufficiently high S (low noise level), an arbitrarily small error probability is attainable with moderate expenditure of energy in the forward path, amounting to one quantum per unit of transmitted information; the overall energy requirements for both forward and feedback paths ensuring similar results come to 3.15 quanta.

4. Systems with interrogation are usually the most efficient when the cost of energy in the feedback path is low and there are strong fluctuations in the intensity of the received signal. Nevertheless, they may also be used to advantage when energy in the feedback path costs more than in the forward path.

REFERENCES

1. T. H. MAIMAN, "Stimulated Optical Radiation in Ruby Masers," *Nature*, **187**: 493–494 (August 1960).
2. A. JAVAN, W. R. BENNETT, JR. and D. R. HERRIOTT, "Population Inversion and Continuous Optical Maser Oscillation in a Gas Discharge Containing a Helium-Neon Mixture," *Phys. Rev. Lett.* **6**: 106–110 (February 1961).
3. J. E. GEUSIC, W. B. BRIDGES and J. I. PANKOVE, "Coherent Optical Sources for Communications," *Proceedings of the IEEE* **58**: 1419–1439 (October 1970).
4. F. S. CHEN, "Modulators for Optical Communications," *Proceedings of the IEEE* **58**: 1440–1457 (October 1970).
5. L. G. KAZOVSKY, "Optimization of the Frequency Characteristic of a Broadband Electro-Optic Travelling-Wave Modulator," *Radio Engineering and Electronic Physics* **15**: 1427–1430 (August 1970).
6. H. MELCHIOR, M. B. FISHER and F. R. ARAMS, "Photodetectors for Optical Communication Systems," *Proceedings of the IEEE* **58**: 1466–1486 (October 1970).
7. D. J. LEVERENZ and O. L. GADDY, "Subnanosecond Gating Properties of the Dynamic Cross-Field Photomultiplier," *Proceedings of the IEEE* **58**: 1487–1490 (October 1970).
8. C. M. McINTYRE, W. N. PETERS, C. CHI and H. F. WISCHNIA, "Optical Components and Technology in Laser Space Communication Systems," *Proceedings of the IEEE* **58**: 1491–1503 (October 1970).
9. J. E. GOELL and R. D. STANDLEY, "Integrated Optical Circuits," *Proceedings of the IEEE* **58**: 1504–1512 (October 1970).
10. W. K. PRATT, *Laser Communication Systems*, Wiley, New York, 1969.
11. A. G. SHEREMETIEV, *Statistical Theory of Laser Communications*, Svyaz, Moscow, 1971 (Russian).

12. R. S. Lawrence and J. W. Strohbehn, "A Survey of Clear-Air Propagation Effects Relevant to Optical Communications," *Proceedings of the IEEE* **58**: 1523–1545 (October 1970).
13. F. E. Goodwin, "A Review of Operational Laser Communication Systems," *Proceedings of the IEEE* **58**: 1746–1752 (October 1970).
14. L. U. Dworkin and M. Schwartz, "The Application of Information Feedback to an Amplitude-Modulated Laser Communications System," *IEEE Trans. Commun. Technol.* **COM-19**, No. 5, p. 618, October 1971.
15. L. G. Kazovsky, "Sequential Analysis under Conditions of Quantum Noise," *Radio Engineering and Electronic Physics* **17**: 1457–1465 (September 1972).
16. L. G. Kazovsky, "Effectiveness of Use of Feedback in Information Transmission in the Optical Wave Range," *Radio Engineering and Electronic Physics* **18**: 1687–1692 (November 1973).
17. A. Wald, *Sequential Analysis*, Wiley, New York, 1947.
18. E. L. Blokh, *Noise Immunity of Communication Systems with Interrogation Links*, Academy of Sciences of the USSR, 1963. (Russian)
19. A. V. Smirnov, *Introduction to Optical Electronics*
20. A. L. Mikaelyan, M. L. Mikaelyan and Y. T. Gurkov, *Optical Generators on Solid State*, Izd. Sovetskoe Radio, Moscow, 1967. (Russian)
21. V. N. Chernyshev, A. G. Sheremetiev and V. V. Kobzey, *Lazers in Communication Systems*, Svyaz, Moscow, 1966. (Russian)
22. V. A. Rozhansky, "Consideration of Multi-frequency of Laser Radiation in Computing Optical Communication Systems," Report at TsNIIS Conference, Moscow, 1968. (Russian)
23. H. Kogelnik, "Modes in Optical Resonator," in *Lasers*, vol. 1 (A. K. Levine, ed.), pp. 373–423, New York, Dekker, 1966.
24. T. G. Polanyi and I. Tobias, "The Frequency Stabilization of Gas Lazers," in *Lasers*, vol. 2 (A. K. Levine, ed.), New York, Dekker, 1966.
25. E. S. Vartanyan, Thesis, Yerevan University, 1971. (Russian)
26. L. G. Kazovsky, Thesis, Leningrad Electrotechnical Communications Institute, 1972. (Russian)

27. J. F. NYE, *Physical Properties of Crystals*, Clarendom Press, Oxford, 1957.
28. W. H. LOUISELL, *Radiation and Noise in Quantum Electronics*, McGraw-Hill, 1964.
29. T. L. PAOLI and J. E. RIPPER, "Direct Modulation of Semiconductor Lasers," *Proceedings of the IEEE* **58**: 1457–1465 (October 1970).
30. M. ROSS, *Lasers Receivers, Devices, Techniques, Systems*, Wiley, New York, 1966.
31. P. W. KRUSE et al., *Elements of Infrared Technology*, Wiley, New York, 1962.
32. C. W. HELSTROM, J. W. S. LIU and J. P. GORDON, "Quantum-Mechanical Communication Theory," *Proceedings of the IEEE* **58**: 1578–1598 (October 1970).
33. S. KARP, E. L. O'NEILL and R. M. GAGLIARDI, "Communication Theory for the Free-Space Optical Channel," *Proceedings of the IEEE* **58**: 1611–1626 (October 1970).
34. D. GLOGE, "Optical Waveguide Transmission," *Proceedings of the IEEE* **58**: 1513–1522 (October 1970).
35. G. L. TURIN, *Notes on Digital Communication*, Van Nostrand Reinhold, 1969.
36. A. A. KHARKEVICH, *Information Theory. Pattern-Recognition. Selected Works*, vol. 3, Nauka, Moscow, 1973. (Russian)
37. D. D. KLOVSKY, *Signal Transmission Theory*, Svyaz, Moscow, 1973. (Russian)
38. K. N. SCHELKUNOV and L. G. KAZOVSKY, "Noise Immunity of Optical Communication Links with Frequency Multiplexing," *Telecommunications and Radio Engineering* **26**: 67–69 (March 1972).
39. B. R. LEVIN, *Theoretical Foundations of Statistical Radio-Engineering*, Izd. Sovetskoe Radio, Moscow, vol. 1 — 1974, vol. 2 — 1975. (Russian)
40. YU. A. KOLOSOV, "Comparison of the Sensitivities of Optimal Optical-Region Receivers," *Radio Engineering and Electronic Physics* **15**: 1430 (August 1970).
41. A. E. BATARINOV and B. S. FLEISHMAN, *Methods of Statistical Sequential Analysis and Their Applications*, Izd. Sovetskoe Radio, 1962. (Russian)
42. E. S. VENTTSEL, *Probability Theory*, Izd. GIFM, 1962. (Russian)
43. V. F. KRAPIVIN, *Wald Distribution Tables*, Nauka, Moscow, 1965. (Russian).

44. V. I. TATARSKY, *Wave Propagation in Turbulent Atmosphere*, Nauka, Moscow, 1967. (Russian)
45. L. M. FINK, *Theory of Discrete Message Transmission*, Izd. Sovetskoe Radio, Moscow, 1970. (Russian)
46. "CO Laser Is Built Small Enough to Fly," *International Electronics* **48**, No. 6, p. 56, March 20, 1975.
47. "French, UK Firms Market Short-Range Fiber-Optic Links," *International Electronics* **48**, No. 6, p. 56, March 20, 1975.
48. "Fiber-Optic Field Attracting New Cable, Connectors," *International Electronics* **48**, No. 10, pp. 33–34, May 15, 1975.
49. "Laser System Runs at 1.5 Mb/s," *International Electronics* **48**, No. 6, p. 169, March 20, 1975.
50. J. E. HARRY, *Industrial Lasers and Their Applications*, McGraw-Hill, 1975.
51. "Optical Fiber Link Transmits 6.3 Mb/s," *International Electronics* **48**, No. 13, June 26, 1975.
52. R. M. GAGLIARDI and S. KARP, *Optical Communications*, 1976.

INDEX

Acceptance region 60, 75, 78
Active medium 2
Active spacing 89
Adaptation 59
Asymptotic behavior 81
Atmospheric channels 46
Avalanche photodiode 31

Bessel function 93

Cathode 20
Characteristic function 71
Cost of energy 110
Crystals 10
Cumulant function 71

Dark current 19, 29
Decision-making time 87, 92
Decision-making scheme 108
Depth of modulation 16
Distribution function 87
Doppler broadening 3
Duplex system 56
Dynode 20

Efficiency 84
   of sequential analysis 84
Electrooptical effect 10
   longitudinal 13
   transverse 13
Energy efficiency 99, 105
Equivalent energy 109, 110
Excitation 1

Fabry–Perot etalon 3
False alarm 78
Feedback 55, 65
   center-of-gravity 59
   decision 58
   information 58
   postdecision 56
   predecision 56
Feedback path 56, 61, 106, 110
Fermi's "golden rule" 42
Fiber optic links 39
Filtration 34
   frequency 34
   polarization 34
   spatial 35
   temporal 35
First moment functions 68, 77
Frequency modulation 68

Gas laser 4
Geometrical interpretation 80
Gram–Charlier series 70, 71
Grouping 76

Hermite polynomials 70
Hypothesis 90

Intensity fluctuations 47
Internal amplification 19
Internal noise 19
Interrogation 59, 75, 76
   symbol-by-symbol 65
Inversion 2

Kerr effect 10

Laguerre polynomials 44
Laser 1
Likelihood function 68
Likelihood ratio 69, 78
Local oscillator 66
Log-normal distribution 47
Lost pulse 78

Memory 59
Memoryless system 94, 98, 99, 102
Mean analysis time 88
Modulation characteristic 15
Modulator 9
  acoustooptic 16
  frequency 12
  magnetooptic 16
  phase 12
  polarization 12
  resonant 14
  travelling wave 14
Modulator nonlinearity 15
Moments 71
Multiplexing 67, 74

Noise 28
  background 29
  internal 19
  multiplication 29
  preamplifier 29
  quantum 28
  thermal 29
  variance of 30
Noise immunity 23, 65
Noiseproof encoding 55
Nonadaptive system 103
Nonlinear distortions 16
Normal modes 12
Number of repetitions 62, 106

Observation period 95
Optical carriers 67
Optical heterodyning 66

Optical indicatrix 10
Optical modulator 9
Oscillation frequency 3

Parallel-channel transmission 55
Parametric representation 98
Passive spacing 77
Payoff matrix 109
PCM (pulse-code modulation) 67
Phase modulation 68
Photodetectors 17
Photoelectromagnetic effect 18
Photoemission effect 17
Photomultiplier 21
Photon-electron transition 42
Photoresistors 18
Phototube, TW 21
Photovoltaic effect 18
Planck's constant 2, 19
Pockels effect 10
Poisson distribution 43, 44, 45, 70, 76, 80
Population 2
Probability 69
  error 70, 72
  posterior 69
  prior 69
Pumping 1

Quantum efficiency 19

Resonator 2

"Satellite/Earth" communication 61
Sequential procedure 75, 90, 91, 108
  Wald 75, 76, 78
Signal/noise ratio 30
Simplex system 56
Solid-state laser 7
Spectral density 43
Statistical hypotheses 75
Steady noise 79

Steady signal 79
Structure function 47
Subcarriers 23

Threshold values 96
Turbulent irregularities 101

Uncertainty region (null region, indeterminate region) 60, 75, 78, 98

Wald approximation 79
Wald distribution function 84
Waveguides 14, 36
  film 37
  gas lens 38
  graded index 37
  hollow 36
  iris 37
  solid lens 38
  surface 36